Energy Efficiency

Energy Efficiency
Concepts and Calculations

Daniel M. Martínez
University of Southern Maine, Portland, ME, United States

Ben W. Ebenhack
Marietta College, Marietta, OH, United States

Travis P. Wagner
University of Southern Maine, Portland, ME, United States

ELSEVIER

Elsevier
Radarweg 29, PO Box 211, 1000 AE Amsterdam, Netherlands
The Boulevard, Langford Lane, Kidlington, Oxford OX5 1GB, United Kingdom
50 Hampshire Street, 5th Floor, Cambridge, MA 02139, United States

Notices
Knowledge and best practice in this field are constantly changing. As new research and experience broaden
our understanding, changes in research methods, professional practices, or medical treatment may become
necessary.

Practitioners and researchers must always rely on their own experience and knowledge in evaluating and
using any information, methods, compounds, or experiments described herein. In using such information or
methods they should be mindful of their own safety and the safety of others, including parties for whom
they have a professional responsibility.

To the fullest extent of the law, neither the Publisher nor the authors, contributors, or editors, assume any
liability for any injury and/or damage to persons or property as a matter of products liability, negligence or
otherwise, or from any use or operation of any methods, products, instructions, or ideas contained in the
material herein.

British Library Cataloguing-in-Publication Data
A catalogue record for this book is available from the British Library

Library of Congress Cataloging-in-Publication Data
A catalog record for this book is available from the Library of Congress

ISBN: 978-0-12-812111-5

For Information on all Elsevier publications
visit our website at https://www.elsevier.com/books-and-journals

Publisher: Joe Hayton
Acquisition Editor: Lisa Reading
Editorial Project Manager: Charlotte Kent
Production Project Manager: Anitha Sivaraj
Cover Designer: Mark Rogers

Typeset by MPS Limited, Chennai, India

Contents

Preface

Energy Efficiency: Concepts and Calculations is meant to serve as a useful guide to engineers, analysts, and policy-makers concerned with energy efficiency. The book's chapters are arranged such that the first four chapters introduce the conceptual aspects of energy efficiency, synthesizing the work of several other authors who have written about energy in more general terms as they relate to physical and environmental aspects. The next four chapters provide a more practical and quantitative look at the energy consuming sectors: electric power generation, industry and manufacturing, residential and commercial end use, and transportation. These latter chapters provide guided sample calculations specifically geared toward understanding step efficiency, relying heavily on reporting and energy analyses performed by governmental agencies. The final chapter takes up policy implications related to energy efficiency and efficiency standards.

This book was written under the premise that all world economies depend on massive amounts of primary energy to function. For the majority of wealthy, developed nations, that dependence comes in the form of fossil fuel energy consumption, with a smaller portion from nuclear fuel, hydropower, and "new" renewable energy. That mix of primary energy is processed into highly versatile fuels that are then converted into heat or electricity to satisfy our varied and growing consumption choices. Ultimately, we make these consumption choices to achieve, or maintain, a high quality of life that necessitates continued demand for more primary energy. However, this demand involves an immense waste of energy and nonenergy resources, as well as generates pollution, for which we must make every practical attempt to mitigate. But how do we reduce the waste and pollution that result from our energy consumption needs?

One important way to reduce primary energy consumption and to mitigate negative impact is to use technological improvements in heat and power systems and in end use machines and devices to reduce energy demand, specifically to maximize the efficiency of primary, secondary, and (sometimes) tertiary energy conversions. That is, use technology to provide the same service or product with less energy input, and therefore less pollution, from fuels and electricity. Even the transition toward alternative energy sources will depend on efficiency improvements, as the classic "renewables" are not readily dispatchable. Technologies already exist to reduce energy demand by between 30% and 70%. Moreover, that realized energy savings results in "found money" for efficiency adopters that likely will generate new economic activity. For all of the abovementioned reasons, we argue that a concerted global effort to focus on energy efficiency would result in the quickest, most sustainable path to reduced pollution and energy waste and improved economic competitiveness. It is under this assertion that we write this book.

Chapter 1, Introductory concepts, presents descriptions of the common terms used to understand energy efficiency, along with means to conceptualize primary energy and energy use. Chapter 2, Dealing with energy units, measures, and statistics, reviews basic energy terms, equivalencies, and statistics essential to understanding energy use and energy efficiency. Chapter 3, Primary energy trends, discusses global distributions and trends in energy resources. Chapter 4, Energy-flow analyses and efficiency indicators, examines the subject of energy-flow analyses by visualizing and calculating energy efficiency within the context of energy-conversion devices, processes, and macrolevel systems. Chapter 5, Electric power sector concepts and calculations, provides a description of electricity generation and delivery, including concepts and calculations related to customer demand profiles, and efficiencies of conversion and electrical transmission. Chapter 6, Industrial sector energy efficiency, reviews the major manufacturing subsectors and describes the various types of industrial energy processes, intensities, and their efficiencies. Chapter 7, Transportation sector energy efficiency, examines consumption in common transportation modes, and the energy processes and efficiencies within the most utilized modes: passenger vehicles and freight road transport. Chapter 8, Residential and commercial sector energy efficiency, describes major residential and commercial services and functions, common devices, and the energy processes and efficiencies within the most utilized component: buildings. Finally, Chapter 9, Policy instruments to foster energy efficiency, places Chapters 5−8 in the context of policies and standards related to energy efficiency.

The authors would like to note a stylistic change in references and citations between the first four chapters and the last five. Chapters 1−4 are conceptual in nature, using general information and knowledge common to energy practitioners. An extensive use of inline references would disrupt the flow of the text in these first chapters, so we have merely placed a summary of references at the end of each chapter. In contrast, Chapters 5−8 rely on a great deal of explicit data and analyses drawn from specific sources. Therefore citations are embedded in the text to provide proper attribution. In addition, reference lists are provided at the ends of all chapters, not only to cite sources used but also to provide ample opportunities to pursue further inquiry into the topics.

Finally, the authors would also like to acknowledge our families for their tolerance in the lengthy process of producing this book. They additionally thank Dr. Robert Sanford and Chris Jacobs for contributing time to reviewing chapters and tracking down original data sources.

Introductory concepts

CHAPTER OUTLINE

Energy Efficiency. DOI: https://doi.org/10.1016/B978-0-12-812111-5.00001-9

In this introductory chapter, we present descriptions of the common terms used to understand energy efficiency, along with means to conceptualize primary energy and energy use. We close with an energy efficiency calculation, assessing total or cumulative system efficiency for lighting, powered by different primary energy sources. Key chapter points include:

- distinguishing efficiency from conservation,
- reasons for studying end use and efficiency, and
- understanding the supply and end-use chain.

1.1 DEFINING ENERGY EFFICIENCY

Energy efficiency is, by definition, a measure of the useful work produced per unit of energy used in an energy conversion. To the extent that energy conversions have a thermodynamic cost, efficiency measures the ratio of the energy sought to the total energy put into the system, such as in an energy conversion device or process. The closer the ratio of energy sought to energy input is to one, the more efficient it is. Thus, efficiency ultimately refers to the ability of an energy conversion device or process to successfully transform one energy form into another more useful form, while minimizing any undesired energy conversions that exist due to the laws of thermodynamics, such as low-grade heat losses that can not be used for any useful purpose. Moreover, improved efficiency refers specifically to technical improvements in devices and processes, which reduce any excess input costs while maintaining the same degree of energy service sought, within a measurable timeframe.

Alternatively, others point out that unit energy consumption (also known as specific energy consumption) may sometimes be more useful for understanding energy efficiency, because it specifies the amount of energy needed to produce a certain amount of product or service from a certain device or process. The smaller the amount of energy used to produce a certain quantity of valuable product or service, the more efficient the device or process is. So, whereas energy efficiency is considered dimensionless (energy out divided by energy in), unit energy consumption has dimensions, such as joules of energy needed to produce a ton of steel, or liters of fuel needed to move a vehicle 100 kilometers. To reduce unit energy consumption in a device or process, thereby improving efficiency, that device or process would need to undergo a design or operational change (i.e., to reduce the number of joules to produce a tonne of steel).

1.1.1 ENERGY EFFICIENCY VERSUS CONSERVATION

Efficiency should not be conflated with conservation, as conservation refers not to technical improvements in processes, but to policy decisions and behavioral choices. It may include decisions to employ or invest in improved technologies, representing an overlap with efficiency. Policies can be designed to require development of energy efficiency (e.g., fuel economy standards) and to encourage the development of energy efficiency products (e.g., ENERGY STAR appliances).

Likewise, policies can be crafted to encourage the installation of energy conservation devices using tax credits and rebates while policies can also be used to promote energy conservation such as adopting progressive rate charges for higher energy consumption. Many conservation choices, though, are voluntary behavioral actions to avoid or defer consumption. These choices can range from rather obvious, "painless" choices, such as turning off lights and electronics when not in use, to choices that may reduce benefits, such as turning thermostats down in the winter.

Efficiency and conservation are often mistakenly viewed as synonymous. While they are related, they are also distinct. For instance, improved efficiency can be a means to achieve conservation, but not the other way around. Efficiency is a technical function of the energy input relative to the useful work accomplished. Choosing to eliminate an activity does not improve efficiency, but rather can lead to clear energy savings—conservation.

Conservation refers to measures to reduce energy use. These measures can include austerity choices to reduce energy consuming activities or incentives to employ more efficient technologies to do the same things with less energy input. Conservation involves making choices, which often include decisions to forego some activities or to change how they are done. It is also true that conservation reliably results from higher energy prices. For example, total miles driven decreases when fuel prices increase as people choose not to take as many pleasure trips or shift to public transportation, taxis, or transportation network companies, but such choices are clearly not examples of improved efficiency.

Efficiency and conservation are certainly related, although not identical. Efficiency improvements typically lead to conserving energy (with the exception of the partial role that the so-called rebound-effect plays). The overlap on the Venn diagram in Fig. 1.1 represents the conservation that is realized by efficiency improvements.

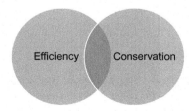

FIGURE 1.1

A Venn diagram of efficiency and conservation. The overlap represents the conservation that is realized by efficiency improvements.

1.2 IMPETUS FOR UNDERSTANDING AND EMPLOYING ENERGY EFFICIENCY

Dramatic transformations in our use of energy lie ahead. Change will be forced on us by ultimate shortages in some of the critical resources on which we have come to depend and on concerns for environmental impacts. Finite resources (coal, oil, gas, and uranium for nuclear fission), which provide the vast majority of the world's current energy sources, will inevitably deplete. Crude oil, the life blood of the world's transport needs and habits, will probably be the first of the finite sources to be constrained by depletion; nevertheless, all the finite resources face these limits. There is also a worldwide desire to reduce emissions from drilling, mining, and mobile and stationary combustion point-sources. Oxides of nitrogen and sulfur, particulate matter, and greenhouse gases are all emissions from increasing consumption of wood and fossil fuel sources for heating, electricity, manufacturing, and transportation, as well as increasing pollution in the natural environment.

Energy efficiency is one of the most valuable responses to these combined challenges. The energy that is not needed, because of enhanced efficiency, will be the equivalent of additional new energy produced. Indeed, it will be better, because efficiency enhancements will generally have little or no on-going environmental impacts. Efficiency improvements are inexhaustible. Once an efficient technology is developed and deployed, it can continue to be used and further deployed until it either saturates the market or is superseded by an even better, more efficient technology. As such, energy efficiency will play an increasingly prominent role in the local, regional, and national agendas of most developed countries.

Of course, there are always other considerations besides efficiency, including pollution and other emissions. For these issues, noncombustion primary sources have a clear advantage. However, evaluating the "life cycle" efficiency of these systems, based on flux-limited resources (e.g., solar and wind), depends on the system boundaries. In particular, does the energy flow being tapped count as energy input, or should we only count anthropogenically controlled energy input? Efficiency must be evaluated within the context of what we seek, which would be the goods and services provided by the energy—at the least direct economic cost.

1.2.1 SOCIAL FACTORS

The ability to harness external energy sources transformed the lives of humans dramatically many thousands of years before anyone developed a language for the economy. Energy systems have continuously evolved to provide greater power for humans and with ever-greater control. We are entirely surrounded by the benefits of energy use: from the simplest forms of cooking meals, to the most sophisticated telecommunications, or the mightiest industries.

Nations using very little external energy have invariably low economic and social advantages, while the affluent nations, with high qualities of life, have relatively high per capita energy consumption (PCEC). Studying the relationship between energy use and quality of life, one finds an exceptionally strong relationship, which in the past we dubbed the "Energy Advantage" and displayed in Fig. 1.2.

That relationship has the characteristics of a saturation curve, meaning that at low levels of energy use and affluence, a little more energy equates to much better quality of life, with the effect diminishing at higher levels, until the saturation level is reached, beyond which greater energy use does not obviously improve quality of life. All affluent, industrial nations are into the flat saturation portion of the Energy Advantage curve, while the United States and Canada are particularly far beyond saturation.

The Energy Advantage is a different kind of efficiency—an efficiency of how much benefit is gained from energy use, rather than of the specific work performed. It suggests that additional energy in the economy has diminishing efficiency, with some energy use in affluent, industrialized nations inefficient in advancing human development.

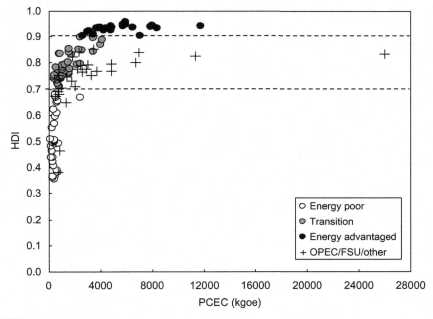

FIGURE 1.2

The Energy Advantage relationship showing PCEC versus the HDI for 120 nations.

HDI, Human development index; *PCEC*, per capita energy consumption.

Adapted from Martínez, D.M., & Ebenhack, B.W. (2008).

This plot of more than 120 nations' human development index (a quality of life indicator deployed by the UN) versus their PCEC shows several important pieces of information in the Energy Advantage. First, there are two distinct trends. The lower trend, shown by the + symbol, represents nations whose economies are energy export dependent. They clearly have different energy consumption dynamics than the rest of the world, but the trends are very similar. The next important information to glean from Fig. 1.2 is the shape of the curve: rising steeply at low energy consumption levels and then leveling off to a horizontal asymptote. This shape is typical of a "saturation curve."[1] The character of this shape illustrates three different regions:

- the energy poor, for whom a little more energy relates to large gains in quality of life,
- the transition, where more energy still adds value, but at decreasing levels, and
- the saturation level, where greater energy consumption does not correlate with additional gains in quality of life.

Legitimate questions can be raised as to cause and effect. Does quality of life depend on the amount of energy supply available or does a higher quality of life cause increased energy consumption? It is important to not conflate correlation with causation; nevertheless, we suggest that there is reason to believe that the availability of energy is essential to support all development activities. Every product and every service we consume is provided by energy. On the other hand, affluence creates demand for all those goods and services that require more energy. It is very likely that there is positive feedback between energy use and quality of life.

1.2.2 MACROECONOMIC FACTORS

Economic factors play an essential role in allocating resources. Energy consumption is promoted by cheap energy. Planning requires some understanding of the economy and what investments will be favored or disfavored under a variety of circumstances. Much of policy-making is about using economic incentives to encourage desired behavior or economic disincentives to discourage undesired behavior.

1.2.2.1 Energy intensity

Energy intensity is a measure that is often used to assess the energy efficiency of a particular economy. The numerical value is traditionally calculated by taking

---

[1]There are many examples of saturation phenomena, in which a little more at low levels makes a big difference, but reach levels at which more has no benefit are oxygen uptake by hemoglobin, or food consumption (if you are starving, a little food makes a big difference, but many of us have passed the saturation point, beyond which more food does not help at all).

the ratio of energy use (or energy supply) to gross domestic product (GDP), indicating how well the economy converts energy into monetary output. Typical units for energy intensity are joules (or Btu) per US dollar; however, there are other equivalent metrics used. The smaller the energy intensity ratio is, the lower the energy intensity of a particular nation. Obviously, low energy intensity is the desired goal, especially within the context of the previous discussion on the Energy Advantage, because it represents an efficient allocation of energy resources to generate wealth and a high quality of life. You are deliberately trying to decouple energy use and economic output to enhance that quality.

It is logical to expect energy intensity to increase, and for GDP and energy use to be closely coupled, during the initial development stages of a nation, since the early years of access to, and utilization of, more abundant modern energy (i.e., fuels and electricity) includes many energy intensive activities. The developing economies will need to produce a tremendous amount of cement for roads and buildings, as well as steel for automobiles, appliances, and telecommunication infrastructures, to name a few. These will be new uses, some of which will naturally be energy intensive. After development has taken hold and industries established, the energy intensity historically falls as cost-cutting is employed to "squeeze" more value out of now established markets and manufacturing subsectors.

Energy intensity can be lowered by several means such as (1) employing advanced energy extraction and conversion techniques, (2) increasing the efficiency of materials production, or (3) allowing other nations to produce the manufactured products and purchase them. These allow for increased GDP via similar or greater product output/usage with the same or lower energy usage, which will subsequently decrease energy intensity. Indeed, there has been a trend of decreasing energy intensities of economies in the latter half of the 20th century, which is expected to continue, due to increased efficiency and to an increase in activities that produce economic value for intrinsically less energy input. Fig. 1.3 shows how primary energy use and GDP have changed globally over time relative to a normalized year of 1971 (the beginnings of the so-called energy crisis era), representing this trend.

Despite the obvious decrease in global energy intensity, it is important also to consider individual, national trends, because many affluent nations have reduced their industrial activity, shifting toward activities which are, by their nature, less energy intensive. This is tantamount to "exporting" the industrial activity to other industrializing nations. Many nations with strong economic growth have also shifted toward more manufacturing, which carries with it more energy consumption, and tighter coupling. To some extent, the "export" of manufacturing to less affluent nations is also a tendency toward less efficient operations that do not necessarily show up in the global assessment.

Regardless, the global trend of diminishing energy intensity is well established and will doubtless continue; however, it can be expected to level out substantially for a couple of reasons. First, as with so many trends, there is a saturation point. Energy conversions are limited and the transition to intrinsically less energy intensive economic activities can only go so far. Economic growth (or even a

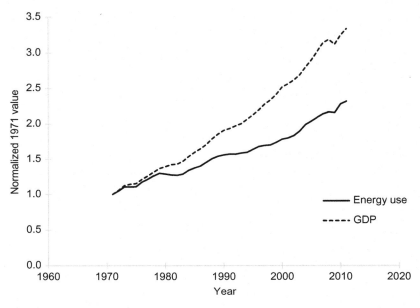

FIGURE 1.3

World energy use and GDP changes from 1971 to 2011, normalized to 1971 values. Decoupling of energy use and GDP has steadily increased (and intensity decreased) with time. *GDP,* Gross domestic product.

Figure created using data presented in US Energy Information Administration International Energy Statistics and World Bank World Development Indicators (2014).

strong steady-state economy) will always require manufacturing: replacing and updating existing devices; building new homes and facilities; providing for transportation; and supporting communications, entertainment, and other services.

1.2.3 OTHER KINDS OF EFFICIENCY

The term "efficiency" can be applied to the use of any resource: time, money, land, water, human labor, etc. It is common to refer to modern agriculture becoming highly efficient. This statement is true in terms of land use, human labor, and economics. It is less true in terms of energy or water resource usage.

There is generally value in the efficient use of any resource—especially resources in limited supply. Food, water, land, and energy are inherently finite resources, which can (and probably will ultimately) face constraints. (It is true that solar-based energy systems are tapping an immensely vast resource base, whose finitude can be ignored for a few billion years, but the flow of even the vast solar resource is finite—and quite limited at any given point on the Earth's surface.)

Although human population is also finite, questions can be raised about the ultimate value of reducing the need for human labor in economic production, until societies approach full employment. "Labor-saving" devices can free people to be more creative and to employ their higher talents more effectively, but there is a point at which humanity does not benefit by needing less human input. Ideally, everyone needs productive employment.

Economic efficiency cannot be ignored. It is a failing of those who argue that what solar photovoltaic (PV) needs in order to gain market share is increased energy efficiency. No. Market share is based primarily on economic efficiency (or perceptions thereof). Most people expect some justification for spending the money on improved efficiency.

Perceptions often fall short of reality, as is seen in the extraordinarily high thresholds often applied to allow investments in energy efficiency. Often cheap, these investments yield good rates of return, but not large dollar payouts. Therefore, many decision makers set exceptionally high rate of return expectations for energy efficiency investments.

1.3 ENERGY SOURCES AND ENERGY CARRIERS

Energy sources are those resources we tap, but the sources are commonly converted into more convenient forms, referred to as "carriers" or as "secondary" energy. Electricity is the most ubiquitous carrier, which offers great and varied utility. Since all conversion processes have some inefficiencies, energy losses are incurred whenever primary energy is converted to a secondary carrier. These losses may be more than compensated for, though, if the carrier incurs fewer efficiency losses than the primary form.

A significant example would be charcoal, which is much lighter and, thus, much less energy intensive to transport, and burns more efficiently than the primary raw biomass form. Charcoal can be considered a carrier for raw biomass. It offers greater portability (due to weight loss of the fuel through dewatering in the charcoalification process), cleaner burn, and greater ease of use. Part of the efficiency gain is that it is much easier to control the heat output and direct it to the tasks, such as cooking. Nevertheless, a significant amount of energy is consumed in the partial combustion process of making charcoal.

Similarly, gasoline, diesel, and jet fuel are carriers for the primary source of crude oil. These secondary products are specifically tailored for the applications. Gasoline, in particular, must meet very tight performance criteria to vaporize for ignition, even in cold starting conditions, while not preigniting during the compression stroke. Conversely, diesel fuel is meant to ignite under the heating of the air in the compression cycle. (The fuel is injected under very high pressure at the top of the compression stroke so that it cannot preignite.) Diesel fuel is composed of heavier, longer chain organic molecules than gasoline. Because diesel fuel does not need to be designed against preignition, its specifications are more broadly defined.

Because of the chemical purity of the refined products, eliminating some of the longest chain, heaviest molecules, the combustion products are much cleaner than burning the raw crude oil. Engines, especially gasoline engines, require very specifically engineered properties so that a portion of the fuel can vaporize (even at very low temperatures) to start an engine and yet not preignite while the air/gasoline mixture is being compressed by the piston, causing knocking of the engine. Since diesel fuel is injected at high pressure at the top of the compression stroke, diesel engines can perform satisfactorily with a much broader range of compositions. Nevertheless, the fuel composition is engineered.

Liquid petroleum gas and compressed natural gas are carriers for natural gas. Natural gas, in its original form, burns very cleanly and efficiently. However, natural gas has extremely low density at normal atmospheric conditions. Therefore, the carriers primarily serve to allow the gas to be transported in ships or automobiles.

Batteries are (secondary) carriers for electricity. There are single use batteries, which commonly hold relatively small amounts of energy as chemical energy, to be released as electricity. These have the advantage of being inexpensive and extremely portable. Multiple use (rechargeable batteries) may also be of small scale, but have the potential for much larger scale, in which they can provide some load balancing benefits, as well. In these cases, excess electricity generated during peak production times (or low demand times) can be stored for use when demand exceeds primary energy production.

1.3.1 PRIMARY ENERGY

Primary energy refers to the sources, which can produce energy that can be controlled by humans. Inevitably, energy is used in bringing the primary energy sources under control, but that energy must be less than the resulting controlled energy in order to be a source. The primary energy sources include the following:

- Combustion fuels
  - Biomass
    - Raw firewood, brush, and dung
    - Charcoal
    - Engineered biofuels
      - Ethanol
      - Biodiesel
      - Waste-stream fuels
      - Biogas
  - Fossil fuels
    - Coal
    - Oil
    - Natural gas
- Nuclear fission fuel (primarily uranium 235)
- Hydropower

- Wind power
- Solar
- Geothermal
- Tidal power
- Wave power
- Ocean thermal

The combustion fuels are grouped together because they are strongly related systems. All of the combustion fuels are based on tapping energy stored in chemical bonds—ultimately through photosynthesis, since the fossil fuels are the remnants of long dead plant life (and some of the animals that ate the plants). In terms of evaluating efficiency of combustion fuels, the most significant distinction is between waste-stream and other fuels. This is because there is no step specific to acquiring (mining, drilling, harvesting) the refuse.

Hydropower is the result of water evaporation by the sun and the vapor being transported in the atmosphere to higher elevations, whereas the rain and snowfall from the headwaters of streams and rivers can be tapped for the kinetic energy of flowing water. Thus, hydropower undergoes two transitions before humans tap it: a phase change from the absorbed heat of sunlight; and potential energy gained as the vapor rises. All of these steps are outside the boundaries of anthropogenic energy systems.

Solar energy is behind most of the energy we tap, leading some authors to claim that solar power is the ultimate source of all energy resources that humanity uses. This is not true. Tidal power is the result of the gravitational attraction of the moon, pulling the earth's surface waters toward it. Nuclear power taps into the energy stored in unstable (fissile) nuclei. Geothermal energy is somewhat related to nuclear power, as it taps the heat of the Earth's interior, which is fueled by the radioactive decay of naturally occurring radioactive nuclei within the Earth. Prevailing geologic theory suggests that the Earth's interior is actually warming due to the decay of these radioisotopes.

Wind is the result of convection currents induced in the air by the sun's uneven heating of the Earth's surface. Heated air rises, which also pushes other air out of the way, creating large-scale convection currents. The energy of waves, in turn, results from wind. Relatively casual observation can show the wind stirring up surface waters.

Biomass growth is the result of photosynthesis using the sun's energy to combine CO_2 and water to form organic molecules. Biomass is the precursor to all fossil fuels as well, which are thus long-stored chemical energy from the sun's light.

Ocean thermal energy is the difference between water heated by the sun at the ocean's surface and deeper cold water. In this technology, rather than heating a fluid and using the low temperatures of surface air or water to condense the vapor (typically steam), the surface temperatures (relatively warm compared to deep ocean waters) vaporize very low boiling point fluids, which drives turbines. The cool deep water temperatures are used to condense the low boiling point fluids. It is much like a heat engine, running in the opposite direction.

1.3.2 **SECONDARY ENERGY**

Secondary energy can also be referred to as energy carriers. They are commonly the result of a first conversion step, from the primary source to a more convenient or useful form.

Electricity is a prominent form of secondary energy, which offers tremendous control. It can be employed to run massive equipment and processes (aluminum smelting plants generally use electricity, and modern locomotives are diesel-electric). There are numerous applications which cannot work without electricity: computing, telecommunications, and scientific instrumentation are a few examples.

Chemical fuel sources are typically converted to secondary forms that have better energy density, cleaner burn, and other desired characteristics. Gasoline, for instance, is very carefully refined so that a significant fraction will readily vaporize, even at low temperatures, to ensure ignition in the cylinders, while not preigniting as the piston heats the air/gasoline mixture during the compression cycle.

Charcoal is a secondary form of biomass, in which the raw biomass is heated by partially burning in a relatively anaerobic environment, so that moisture is driven off. Inevitably some of the chemical energy content is lost as well, but the charcoal product burns more efficiently at the point of consumption and is lighter, thus requiring less energy to transport. It also burns hotter and cleaner.

1.4 **THE ENERGY SUPPLY/DEMAND CHAIN**

An energy system is commonly understood as the juxtaposition of energy supplies and energy demands, which is often sequenced as an energy supply/demand chain of (1) supply acquisition, (2) primary energy conversion and processing, (3) fuel/electricity distribution and transportation, and (4) end-use conversion, utilization (not including post-use waste management).

1.4.1 **SUPPLY ACQUISITION**

Supply acquisition entails the identification and then extraction of energy sources. As we commonly do not use raw energy directly to perform the desired work, we must identify resources where they are located, quantify the amount available at their source location, and then extract them for processing.

1.4.1.1 *Oil and gas exploration and production*

Oil and gas occur in the tiny pore spaces of rocks, commonly found several hundreds of meters below the surface. There is a common misperception that modern geophysical methods can detect oil and gas from the surface. In fact, these methods enable geoscientists to map the subsurface structure quite accurately, but only drilling an exploratory well into the potential reservoir rock can evaluate the quantities and productivity of oil and gas in the pores of those rocks. Of course, some energy is consumed at each step: generating geophysical data across large areas at the surface and drilling the exploratory wells. The overall assessment of the energy efficiency of oil and gas systems must include the energy input to take these preliminary analytical steps, which do not, themselves, result in the production of any useful energy.

1.4.1.2 Exploration for and evaluation of uranium deposits

Nuclear fuels are extracted from rocks with relatively high concentrations of uranium. Several kinds of geologic formations can hold uranium ores. In many cases, test holes must be drilled to assess these subsurface formations.

1.4.1.3 Coal bed exploration and mapping

In the early days, coal resources were exploited where it was seen in outcrops. Since coal is a very soft "rock," it is readily eroded, leaving indentations in the wall of river banks. Thus, the exploratory process proceeded to mapping the direction and extent of these exposed coal beds. Of course, in modern times, deeper coal beds are exploited, so test holes are often drilled to identify and quantify coal beds. Relative to the energy content of coal reserves, the energy losses in exploration are minimal.

1.4.1.4 Wind speed analysis and measurement

The energy that can be harvested from wind is proportional to the cube of the wind velocity. Generalized wind maps exist for much of the world, but before investing in a major wind project, it is important to develop detailed wind maps for the site, showing the average wind speeds, but also the day-to-day and hour-to-hour wind speeds. Wind speed tends to increase with altitude above ground level, as there are fewer obstructions, meaning that the maps must be three dimensional. Indeed, they should be four dimensional, to account for seasonal variations.

1.4.1.5 Solar insolation mapping

The amount of sunshine reaching any given point is referred to as "insolation." Much like wind, insolation maps exist for most of the world. The variations related to day and night are predictable, but seasonal variations of cloud cover must also be detailed, as well as site-specific topography.

1.4.1.6 Mapping river flows

Modern hydropower is based on harnessing the kinetic energy of flowing water to turn dynamos, which generate electricity. The amount of electricity that can be generated is directly proportional to the volume of water flow that can be tapped and the pressure that it exerts. The pressure is, in turn, a function of the height through which the water is falling. The amount of energy available, then, depends on the characteristics of a river's flow. In this case, the identification stage is quite literally a mapping process, characterizing river flows in detail.

1.4.1.7 Geothermal energy identification

Geothermal energy is ubiquitous. Anywhere on Earth, temperatures increase with depth below the surface from the average ambient surface temperature. Indeed, anywhere on Earth, even very shallow ground shields temperatures from the fluctuations of fickle weather. In summer, the ground is cooler than the air on hot days, while in winter, it is warmer. This, alone, offers a heat source and sink from which to draw heat or into which to expel heat. This use of geothermal energy can be very effective, but is limited to a specific end use (heating and cooling).

Geothermal energy can also be tapped to provide steam for large-scale power generation. These applications require high temperatures and are currently restricted to relatively shallow reservoirs, which contain high-temperature water or steam. Understanding of geologic provinces tells us where magma has flowed from its typical shallow depths, nearer the surface, to heat water to steam. It is easy enough to observe volcanic activity, hot springs, and steam fumaroles, indicating the presence of shallow high temperatures. As with oil and gas, though, only drilling wells can identify producible steam reservoirs.

1.4.2 **ENERGY EXTRACTION**

Energy extraction is the tapping or acquisition of energy from a primary source.

1.4.2.1 Oil and gas production

As noted earlier, exploration for oil and gas includes drilling the first well or wells. Drilling is energy intensive, but once the wells are drilled, oil, and gas flow out of the wells with extraordinarily little energy consumption. Generally, after a few years, the pressure in crude oil reservoirs declines to the point that the oil can no longer flow up the wellbore to the surface, requiring pumping or some other form of "artificial lift." Gas wells, though, essentially never require pumping and flow until most of the gas has been extracted from the reservoir. Very little of the net energy produced from oil and gas is consumed in the extraction phase, since oil and gas are even more energy dense than coal.

1.4.2.2 Coal mining

Coal is a solid fuel that is generally extracted through underground mining or mountain top removal and other types of surface "strip" mining. All of these activities are energy intensive. In terms of overall system efficiency, though, the energy consumed in mining is still not a very large factor, because energy is being extracted in such large quantities.

1.4.2.3 Wind and water turbines

Large-scale, modern wind power is based on very large-bladed wind turbines, which may tower 80 meters. These turbines require substantial energy (and material) to manufacture. Because the energy source being tapped is low density and somewhat intermittent, the energy inputs to build the facilities are more significant to the overall energy produced. Very little energy is consumed during the operational life of wind turbines though.

Hydropower, on the other hand, is a relatively dense energy source, because water is nearly 800 times denser than air. Dam-based hydropower stores and concentrates the energy source, making for an intense, dependable energy source. A large dam, though, is extremely energy intensive to build.

Early use of hydropower was less disruptive, tapping the energy of naturally flowing water, by inserting a wheel into a stream, which turned a mill wheel or pump—simply converting the kinetic energy of the flowing water to the kinetic energy of the machinery. There is a return to the "run-of-the-river" technology with modern hydroelectricity, which does not require the energy intensive construction of a dam. In spite of the savings on initial investment, though, the cost-effectiveness is likely to be reduced by the loss of "economy of scale."[2]

1.4.2.4 Harvesting raw biomass

Raw biomass, in the form of firewood, charcoal, and dung, can be acquired in many different ways. Some firewood is simply gleaned from the forest floor. In some cases, it may be gleaned from prunings and waste streams. In the developing nations, rural people (often women and children) commonly walk out to the forests to gather wood. The costs in terms of injuries and other hazards are high, but not the energy input. In urban areas, people are still often firewood dependent, but cannot walk to the forests to gather firewood. Markets develop for fuelwood in which entrepreneurs with vehicles travel out to the forests and return with loads of firewood or charcoal. (Recall that producing charcoal actually represents a conversion to a carrier, with efficiency losses associated with that conversion.) Fuel consumption driving to and from the forests, yields efficiency losses in the biomass market system, especially since firewood has relatively low energy density—so, a great deal of material must be transported to carry much energy to the market. The dynamics of fuelwood harvesting, possible conversion to charcoal, and transport to market are not well understood and would seem to be important topics for study, as fuelwood dominates energy use for the half of humanity who depend on it.

In general, there is little difficulty in identifying the location or occurrence of firewood—you can see the forest for the trees. However, there may be a real challenge in evaluating forest loss. One can speculate that forest loss in firewood-dependent regions may be worse than appears. Forest extent can be estimated by satellite imagery. However, once fuelwood markets develop, there is likely to be a preference for certain tree species that are more convenient to cut and haul, or are preferred in the marketplace. Therefore, other trees and brush may be left behind, appearing to be forest on satellite imagery, in the absence of detailed ground-truth surveys.

---

[2]Economies of scale refer to the reduced unit cost of materials and construction due to purchasing in large quantities and mobilizing equipment to a single central location.

1.4.2.5 Solar power production

Since solar PV systems generate electricity as a carrier directly, we will consider its efficiency issues in the context of primary energy conversion. It tends to run rather low, but it is important to remember that the primary source arrives at Earth's surface free and in abundance. Therefore, the efficiency of the PV conversion does not affect the economic efficiency—only the amount of energy that can be converted at the installation.

A totally different form of solar power production is solar thermal electric or concentrating solar power (CSP). In the CSP case, reflective surfaces are used to concentrate the sunlight into a point or line, where the heat boils a working fluid to generate steam to drive a turbine, similar to steam turbines in a coal-fired power plant.

1.4.2.6 Nuclear power

Uranium is the primary source of nuclear fuel. It must either be mined or extracted through wellbores by "in situ leaching" of uranium minerals from the surrounding rock. When mechanically mined, the uranium is found mixed with other minerals and requires an initial purification process.

Even once pure uranium is obtained, the fissile[3] portion is less than 1% of the pure uranium. The purified uranium then undergoes "enrichment" to separate the fissile isotope. Separating two isotopes of the same chemical materials is very challenging. For uranium, the material is commonly reacted chemically with hydrofluoric acid to make gaseous UF_6, which can be separated in a centrifuge. Then the lighter fraction of UF_6 containing fissile U235 must be chemically separated. These steps make the acquisition of uranium fuel very energy intensive, but the fission reaction is so highly energetic that the overall efficiency lost in acquisition is relatively low.

1.4.2.7 Geothermal production

Geothermal can be utilized in three fundamental ways: (1) extracting heat to make steam to drive a turbine heat engine to generate electricity, (2) using the "reservoir" as a heat sink to extract heat for space heating or to which to reject heat into it for space cooling, and (3) using the heat "passively" for space and water heating.

The first electric power generating technology represents the only large-scale application of geothermal energy. Presently, only high-enthalpy reservoirs

---

[3]Fissile means that it is capable of undergoing fission directly, when bombarded with neutrons. It can then undergo a sustained "chain reaction" as the fission products release more neutrons that strike more fissile nuclei, creating more fission products (including large amounts of energy.) The nonfissile 99% of uranium is "fertile" though. These fertile nuclei can, under proper conditions, absorb neutrons, undergo decay, and transform into heavier "fissile" elements. This would be the basis of breeder reactor technologies.

containing steam or very hot water that can be vaporized or transfer heat to vaporize a secondary working fluid. These resources are limited in occurrence.

The energy input term to evaluate the efficiency of high-enthalpy geothermal energy can be problematic. As for the purely flow-limited resources (e.g., solar and wind), geothermal taps into a natural energy source, rather than a fuel or feedstock from which to produce energy. So, does the input energy include the heat that is not extracted from a geothermal reservoir? It probably should not, as only the energy tapped or extracted by human activity from the natural energy source is input into the system.

In reality, the geologic formations tapped for steam are not finite reservoirs, like oil and gas reservoirs. Rather they are heat exchange systems. The hot water produced as steam is easily replenished by injecting the steam condensate (and perhaps some additional "makeup" water to account for steam losses). The heat is replenished to some extent by heat from all of the surrounding rocks. The efficiency of extracting heat is complicated to evaluate, as there is the original heat in the reservoir and the heat flowing in from all of the surrounding rocks, once the temperature of the reservoir drops.

Finally, geothermal heat pumps serve as a very different kind of technology. They use the earth as both a heat source when space heating is needed and as a heat sink when cooling is needed. This is a very interesting case for resource efficiency, because the heat flow goes in both directions. During the summer, heat is being injected into the ground. At least theoretically, this replaces heat that is extracted for space heating in the winter. In practical terms, neither of these factors is likely to be significant in the big picture, as the heat flux is tiny compared to the heat capacity of the ground. As with other heat exchange systems, a coefficient of performance is used to evaluate how effective or the heat pump can extract or insert energy within the ground-source water.

1.4.3 ENERGY CONVERSION

In general, energy conversion is the initial process involved in converting one energy form into a more useful form. Frequently, the source is converted to a carrier that is (1) more convenient to use, (2) easier to store or transport, or (3) more versatile. Crude oil is typically converted into gasoline and other fuels that are energy dense and burn efficiently in highly specialized engines. Electricity is one of the major energy carriers in the world, as it offers tremendous versatility. It is at the heart of a range of appliances and devices that provide services unimaginable a few decades ago.

1.4.3.1 Thermomechanical conversion for electricity generation

A vast majority of the electric power generated in the world is from combustion fuels (primarily coal and natural gas). The process involves generating very hot gasses, whose kinetic energies turn turbines (fan-like sets of blades), which spin dynamos to generate an electric current. These processes have theoretical upper limits defined by what is known as the Carnot efficiency: which is a function of the ratio between the inlet and outlet temperature. The hot gasses may be generated by burning fuel to boil water or running the combustion gasses directly through a turbine (as in natural gas turbine technologies). These processes are seldom used in an initial conversion stage, such as generating electricity, thus belong properly in the end-use conversion category.

1.4.3.2 Electromechanical energy conversion for electricity generation

Electromechanical energy conversion is similar to the second step of the thermomechanical processes described above for electricity generation. Rather than having to generate steam to push the turbine blades, the preexisting movement of wind or water drive the turbine. The energy conversion for water-powered systems is quite efficient, as it is a single-step conversion and water is a very dense fluid, so it can impart a great deal of energy to the turbines. Note that for flow-based resources like wind, solar, hydro, and geothermal, extraction and conversion happen simultaneously. These systems also are based on natural flows and stocks of energy, for which the first intervention is conversion (generally to electricity), rather than to mine, pump, plow, or harvest the source.

1.4.3.3 Refining oil into energy (and nonenergy) products

Crude oil is a wide-ranging set of mixtures of organic molecules, called "hydrocarbons" because they contain only carbon and hydrogen—but in a large range of molecular sizes and arrangements. The crude oil constituents can be separated and, through a variety of reactions, reconfigured into a far greater array of chemical products than in the original crude oil. The refinery products include fuels with specific properties for gasoline, diesel, jet fuel, and heating oil. Energy is consumed in all this processing.

Gasoline, diesel, and other fuels can be viewed as energy carriers, offering cleaner, better controlled end-use conversions. We will, however, not treat this is as a conversion process to examine. In addition to fuel production, petroleum is used as raw feedstock, which provides an amazing and often overlooked array of synthetic compounds that provide us with fabrics, plastics, pharmaceuticals, and many more. Do the values of these nonenergy products figure into the overall efficiency of petroleum? Some will argue for including by-products in efficiency calculations for other energy systems, such as corn ethanol. This question certainly complicates the effort to understand the overall efficiency of energy systems.

1.4.3.4 Chemical conversion of agricultural crops into fuel alcohol or biodiesel

The crop-based fuels do not require any exploration to find—rather conversion of other agricultural lands or natural productive lands (also known as arable land) into production of the chosen biofuel crop. This initial land conversions is likely to have environmental costs and some energy costs as well. Acquisition of the fuel requires tilling and fertilizing the soil, planting, pest control, irrigating, harvesting, and transporting the crop to a facility for initial conversion to a carrier fuel. All of these steps are energy intensive, spurring debate about whether the biofuels actually have a positive net efficiency.

1.4.4 ENERGY PROCESSING

Energy processing involves converting energy into a secondary, carrier form. Energy carriers are what most of us generally experience, due to their convenience, energy density, and, often, cleaner characteristics, when compared to the original source.

1.4.4.1 Electricity

Probably the most important and pervasive energy carrier is electricity. Its versatility is unmatched, as seen in the array of appliances and devices powered by electricity. The energy services that we take for granted in modern affluent societies would be inconceivable just a few generations ago. At the point of consumption, there are no emissions. It is clean and quiet. (This sometimes leads people to ignore the impacts of the overall system.) Electricity can be controlled with extraordinary precision, enabling its use in the most delicate applications. It can be transmitted over any range of distances. It can be carried with the energy density of high voltages, or reduced to microvoltage. It cannot, however, be stored as electricity.

1.4.4.2 Batteries

A common means of storing electrical energy is in batteries. This process uses the electricity to create charge separation in a container, that is, the battery cell. There are both inexpensive single use batteries and much more expensive rechargeable varieties. The efficiency of recharging batteries tends to be good, but considerable energy is used in manufacturing batteries of either variety, relative to the small amount of energy they store. The reason that small batteries can seem effective is the high efficiencies of many modern electronic devices.

1.4.4.3 Hydrogen

Although fuel cells have not yet made the commercial inroads that some predicted at the beginning of this century, there is potential for hydrogen to become a significant energy carrier. Hydrogen naturally exists in a diatomic molecular state. When chemically reacted with oxygen, it produces water and energy. A hydrogen storage/fuel cell system works very much like a battery. In this case, it prevents the hydrogen and oxygen from reacting directly, but separating the hydrogen nuclei from their electrons and providing an electrical circuit for the electrons to flow through (thus producing electricity), while the hydrogen nuclei may flow through a special membrane to react with the oxygen. The process is appealing, because it is very clean, but the efficiency of generating the pure hydrogen and storing it until it is needed is problematic.

The standard storage envisioned at this time for hydrogen fuel cell vehicles is to compress the gas to more than 10,000 pounds per square inch—a pressure equivalent to the weight of two sport utility vehicles on every square inch. This raises questions of safety as well as energy efficiency. In addition to the engineering design for safety of the storage container, there is the question of transferring hydrogen to individual storage containers.

1.4.4.4 Gasoline

The predominant product derived from crude oil is gasoline. This is not because gasoline would naturally be the primary product, but because it is in such high demand. This is especially true in the United States, where individual vehicular use dominates the transportation sector. In order to meet the large demand for gasoline, a significant amount of refinery technology is focused on expanding that product—at inevitable efficiency costs. The need to implement additional separation and synthesis processes to increase the gasoline yield has efficiency costs. Just as in every conversion, each step of a process has associated energy losses.

1.4.4.5 Diesel and heating oil

Diesel and heating oil are both drawn from essentially the same distillation column runs. These products require less refining and have higher energy densities than gasoline. Thus, they have fewer efficiency losses in the refining stage. (Diesel engines tend to operate more efficiently than gasoline engines at the point of end-use conversion.) Since diesel can operate with a broader chemical range than gasoline and does not depend on a string of exotic synthesis processes, less energy (and less efficiency losses) goes into the refining of diesel fuel compared to gasoline.

1.4.4.6 Jet fuel

Jet fuels are a class or grade of a more general aviation fuels category. Most precisely, jet fuels are designed to burn in varieties of engines which spin turbines directly from the expansion of the combustion gasses. The other form of aviation fuels are intended for use in reciprocating (piston) engines—more like automobile engines, but driving propellers rather than turning wheels through a drive train. It is derived largely from blending runs from the refining distillation columns. Its boiling point is intermediate between gasoline and diesel fuels.

1.4.4.7 Wood fuel (chips, pellets, and/or sawdust)

Processed wood fuels have historically been based on waste streams from lumber and other operations. This raises interesting questions about how to evaluate efficiency, since the energy content of these waste streams results from producing other goods. Nevertheless, it is possible to consider the starting point as the calorific energy content of those waste-stream materials, as they are processed and prepared for later consumption. In recent years, however, fast-growing tree crops are being utilized exclusively to generate wood chip and/or sawdust fuel to be blended with coal. In some instances, these fuels are being grown in different countries separated by entire oceans. This could have huge impacts on efficiency if adopted on a wide scale.

1.4.4.8 Charcoal

In fuelwood-dependent regions, charcoal is a preferred energy carrier, both because it burns more cleanly, and because it is easier to handle and control. Unfortunately, a common technique for making charcoal in lower income countries is to fell a tree, start it on fire, and partially smother the fire with dirt. This is an extraordinarily inefficient process. However, the charcoal product is lighter to transport and burns more efficiently than raw firewood. It can be difficult to assess the overall efficiency compared to burning the firewood directly.

1.4.4.9 Engineered biofuels

Ethanol produced from corn or sugarcane feedstocks and biodiesel products (such as from soy) are characterized here as "engineered" biofuels. These are the fuels that require specific processes to produce a uniform fuel that can perform well in engines: primarily alcohols and diesel fuels. When a fuel is produced from crops grown specifically to produce that fuel, all of the energy required to grow, harvest, and process count as energy inputs in the calculation of the energy system efficiency. This means that most biofuel crops have very low overall efficiencies. In terms of comparing the values of fuels, it is important to also consider the inefficiencies of utilizing other resources, including water and arable land.

Some authors claim that corn ethanol has a negative energy balance (efficiency). This means that more energy goes into the sum of all of those processes than the fuel contains. If this is true, then ethanol derived from corn cannot truly be considered an energy source—only an energy carrier. Those who argue, conversely, for a positive corn ethanol energy balance base their support on two factors: counting the "energy value" of by-products produced along with the corn ethanol[4]; and often arguing only to count the petroleum-based energy inputs. Both of these approaches are problematic. While it may make sense to count the value of by-products as part of the system's "useful work" if this approach is applied to any fuel, it should be comparably applied to any competing alternatives to which the fuel is being compared. It can be challenging to asses all of the by-products of every energy system fairly. Certainly, the second approach, of only counting energy inputs from petroleum, defies the basis of evaluating energy efficiency. It presumes that the superiority of biofuels over petroleum is a given, but is clearly not an energy efficiency calculation.

---

[4]The chief by-product is "dry distiller's grains" that can be used as livestock feed. The ethanol proponents, then, argue that these by-products are special for their fuel. Some researchers have suggested that gasoline lacks such by-products. This is an especially problematic claim, in light of the vast array of high-value petrochemical feedstocks that are produced as by-products of petroleum refining.

Biofuels derived from waste streams do not have to consider the energy inputs for growing and harvesting the crop. The crop is already being grown for other purposes. This is what makes ethanol from cellulosic feedstock something of a "brass ring" in terms of efficient biofuels. The corn stalks that make up the cellulosic feedstock are, at least in part, waste from corn agriculture. Much of the feedstock for ethanol from sugarcane is bagasse, which is a secondary by-product in sugar extraction.

1.4.5 ENERGY DISTRIBUTION/TRANSPORT

The energy acquired commonly must to be transported from the point of original production to the point of primary conversion (e.g., shipping oil from the wells to the refinery). Ultimately, it also needs to be delivered to the consumer or point of end use. Whether shipping oil, piping gas, transmitting electricity, or hauling coal, energy is consumed in the process, which detracts from the overall system efficiency.

1.4.5.1 Natural gas pipelines

Pipelines are the most efficient means to move mass, except for large, long haul maritime ships. In affluent nations that have access to natural gas, pipelines bring the product all the way from the well to the consumer. Unlike large ships, pipelines are highly scalable, operating anywhere from large transcontinental pipelines, all the way down to small gas delivery lines to homes. Gas pipelines, in particular, are capable of delivering fuel to the point of consumption, even in quantities sufficiently small and precise to simmer a small pot on a stovetop.

1.4.5.2 The electric power grid

The grid that distributes and delivers electricity has become a series of extensively interconnected networks. In some ways, they resemble the circulatory systems of living creatures. Large regional transmission lines branch into smaller local transmission lines, which subsequently branch into smaller lines delivering electricity to each consumer, like major arteries branching into smaller ones, which branch into capillaries to feed each cell in bodies. These extensive electric power grids offer a number of advantages. They permit moving electricity long distances with the security of being able to shift electricity to areas in which there are grid failures or demand spikes. The extensive grids are essential to distributing centralized power generation to large and widespread population centers. They also have been seen to create vulnerabilities when a failure in one part of the grid can cause cascading failures bringing down more of the grid.[5] They require massive investments in infrastructure and its maintenance. Some

---

[5]Think of the ropes that mountain climbers use to tie themselves to one another: the ropes allow other climbers to support one who slips, but if a fall is sufficiently severe, the falling climber may, in fact, pull others down with them.

alternative energy proposals call for centralized power production in locations with consistently high winds or arid conditions (to maximize the electricity produced from relatively uninterrupted daytime insolation). High unrelenting winds and deserts are not appealing population centers. Therefore, these solar and wind farm proposals are likely to require longer major transmission lines, with associated efficiency losses that may impair the efficacy of these systems, in order to move the electricity from the point of production to the consumption centers. For this reason, it is more likely that distributed power projects will be favored for the foreseeable future.

1.4.5.3 Hauling coal by rail or barge

Barge and then rail transport were instrumental in enhancing mobility and promoting expansion into open spaces. They are both relatively efficient transport modes. Barge use is limited to waterways, but that problem was addressed by digging canals—of course, they still require relatively flat land. Rail transport is somewhat less efficient, but can be faster, and rail is easier to lay than canals are to dig. Rail transport is also considerably faster than barge transport, making it a more efficient use of the time resource. The advent of extensive rail systems stunted the expansion of canal systems. Because coal is solid and massive, it is not readily suited to pipeline transport.[6] A large portion of coal is hauled by trains. Large-scale power plants commonly have contracts with mines for unit trains, which exclusively travel back and forth between the mine and the power plant. This usually consists of 100 cars, each carrying approximately 100 tonnes. That is a great deal of mass, being hauled overland to produce power. The trains are likely to return to the mine without cargo, "deadhead," which uses more energy without producing any additional desired useful work. Another large share of coal is transported by barge. This is relatively efficient but is limited by the presence of navigable rivers or canals.

1.4.5.4 Trucking fuel to service station or homes

Generally, trucking is reserved for relatively short distances but still incurs larger efficiency losses than the much more efficient long distance pipeline. Trucking is the least efficient of the energy transport modes, but is the most versatile, being able to move modest amounts to a wide range of locations without having to build dedicated infrastructure.

1.4.6 ENERGY END-USE AND UTILIZATION

At the point of end use, the delivered energy undergoes a final conversion to produce useful work. This may be driving a car, powering an appliance, cooking a meal, or many other kinds of work. Most of the attention to energy efficiency has

---

[6]Coal can be pulverized and mixed into a water slurry for pipeline transport, but this technique has not gained broad application, partly over concerns about the inefficiencies of water resource used.

tended to focus on the final step. Energy end-use utilization is the stage of finally producing useful work to the consumer. It typically entails end-use conversion of a delivered fuel/electricity to satisfy a basic human need or service. Examples include powering electronics, converting electricity into light or heat, or powering an engine.

1.4.6.1 End use

End use is the point at which the energy is ultimately converted to useful work. The end-use technologies are those which perform those final conversions, whether a vehicle for transportation, an entertainment or communications device, a home appliance, a heating and cooling system, a light, or machinery at a factory. End users are those who ultimately consume the energy for useful work. Since the end-use is the most visible to the consumer, these technologies are also highly visible and the most likely for consumers to consider their efficiencies.

1.4.6.2 Burning fuels for space or process heat

Space heating is essential to survival, let alone comfort, in climates that experience temperatures below about 10°C (50°F). Process heat includes systems that use energy for heat to carry out a process, whether it is a blast furnace, a distillation column, or processing food (cooking it).

Natural gas is the most efficient energy source for space heating applications, often approaching 90% conversion efficiency at the point of combustion. With no primary conversion required[7] and only a few percent losses in acquisition and delivery, the total efficiency can exceed 80%. Electric resistance heating enjoys 100% efficiency at the point of use, but typically less than 40% primary conversion for coal-based electric power systems (in addition to minor efficiency losses in mining and transport). Heat pumps, although sometimes considered more than 100% efficient at the point of use, due to tapping non-anthropogenically controlled heat from the surroundings, the total system efficiency, which includes fuel consumption at the power plant, is still not much greater than 50%.

1.4.6.3 Powering electronics

Electronics have become much more efficient power consumers in the early 21st century, in part due to the prevalence of portable, battery-powered devices. In order for the batteries to operate devices for satisfyingly long times on each charge, they must not waste any more power than necessary. The efficiency improvements have been adopted by stationary applications as well. Once an efficient technology is proven to be effective and cost-effective for portable devices, why not adopt it in similar, non-portable applications as well?

There has been such a remarkable proliferation of electronic devices throughout societies that, without significant efficiency gains, growth in electric power

---

[7]Natural gas tends to be very pure mixtures of light alkanes (straight chain organic molecules), with relatively few occurrences of contaminants that need to be removed.

demand would have been extraordinary. Consider the array of electronic devices in contemporary society: laptops and tablets, mobile/smart phones, and all kinds of power tools, including the myriad cordless tools with rechargeable batteries.

The improvements in operating efficiency are largely offset by the standby "phantom" loads. A tremendous number of electronic devices either have clocks and status lights, or they maintain some current (even when turned off) to enable being powered up quickly. These features cause many electronic devices to consume electricity constantly. If we consider the useful work done to be the function of the electronic device, then the phantom power draw is wasted, thereby reducing the overall system efficiency.

1.4.6.4 Converting electricity into light or heat

Two of the most critical uses of energy in modern societies are lighting and providing heat. Heat may be used for cooking or for space heat. Lighting devices tend to be very inefficient, while heating applications tend to be quite efficient. Light is a more challenging application, as only specific wavelengths are appropriate for lighting our living spaces—and the light may scatter to areas that are not the target of the lighting task. A great deal of the electricity ends up being dissipated as heat, which can be viewed as a lower grade of energy than white light.

The point of heating, on the other hand, is to produce lower grade energy. There is no waste in electricity that is converted to heat in this case. Space heating is considered to be 100% efficient, because all of the electricity is converted to heat in a resistance heater. Cooking is not as efficient, but, with electric or natural gas stoves, the heat can be targeted very precisely on the pot or pan. Perhaps even more importantly, the heat can be turned on and off promptly at the beginning and end of the task. On the other hand, biomass fuels (firewood and charcoal) must be started burning before the meal is cooked and cannot be readily turned off when the cooking is done, resulting in waste.

1.4.6.5 Powering an engine

Most vehicular and equipment engines are powered by internal combustion. These engines commonly use gasoline or diesel fuels, but some can use fuel—alcohol mixtures. At the point of consumption, diesel engines tend to be more efficient than gasoline, largely because diesel engines can operate at significantly greater compression ratios than gasoline.

1.5 EFFICIENCY CALCULATION: ASSESSING CUMULATIVE OR TOTAL SYSTEM EFFICIENCY

Let us finish this chapter by considering the flow of energy through systems. Energy is originally tapped by human activity, but a sequence of steps begins, with some energy losses incurred at each step. Some energy systems have many steps, while others have very few. Moreover, there are large efficiency losses at some conversion steps, while some are minimal. By conceptualizing efficiency as a system of steps, we can then formulate an expression that calculates total system or cumulative efficiency:

$$\text{Cumulative Efficiency} = \text{Efficiency of Step 1} \times \text{Efficiency of Step 2} \times \cdots$$
$$\times \text{Efficiency of Step } N$$

This simple algebraic expression illustrates the accumulation of losses for any defined system and highlights that each system is controlled by the lowest efficiency step (or steps), affecting overall system efficiency.

In the following examples, we will consider the stepwise efficiency losses for various processes of producing illumination. Each process has one or more steps with sufficiently low efficiencies as to be a controlling step. We have chosen examples that demonstrate large differences in efficiency and have also considered how we treat energy sources in efficiency calculations.

1.5.1 SYSTEM EFFICIENCY OF ILLUMINATION VIA INCANDESCENT LIGHTING FROM COAL-FIRED ELECTRICITY

Let us start with an extremely inefficient process of generating electricity in a conventional coal-fired steam power plant to power an incandescent light bulb in your home. Starting with coal fuel that will be mined, we can estimate the total or cumulative efficiency in the following fashion:

Step 0: Coal that will be mined (no system processes)
   Step efficiency = 1.0; cumulative efficiency = 1.0
Step 1: Extraction and delivery of coal (roughly 10% loss in extraction and delivery)
   Step efficiency = 0.9; cumulative efficiency = 1.0 × 0.9 = 0.9
Step 2: Energy converted to electricity (roughly 65% losses in coal power conversion)
   Step efficiency = 0.35; cumulative efficiency = 0.9 × 0.35 = 0.315
Step 3: Transmission of electricity (roughly 10% loss in energy delivered to consumer)
   Step efficiency = 0.9; cumulative efficiency = 0.315 × 0.9 = 0.2835
Step 4: Incandescent lighting (roughly 98% losses in incandescent lamp conversion)
   Step efficiency = 0.02; cumulative efficiency = 0.2835 × 0.02 = 0.00567

Based on these estimates of conversion losses, the total cumulative efficiency of converting the chemical energy in the coal acquired at the mine into light using an incandescent bulb is equal to 0.00567 or 0.567%. This appears to be an extremely inefficient process.

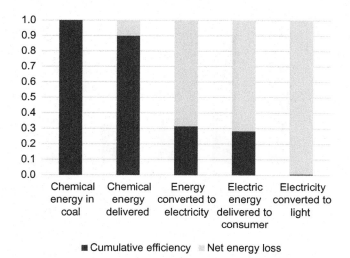

FIGURE 1.4

Efficiency losses by step from coal to incandescent light.

To have a visual understanding of the energy flows described above (i.e., from the raw chemical energy content of coal to illumination provided by an incandescent light bulb), consider Fig. 1.4, which illustrates the stepwise losses just detailed.

The light gray portion of the bars totalizes the losses from the original chemical energy to the current step. They demonstrate how little of the original energy that we tap is ultimately carried through to produce the desired useful work. In this particular case, there are two very inefficient steps: converting coal to electricity and converting electricity into light. By far, the least efficient step is the final conversion to light. Therefore, this is the step that offers the greatest potential for improving the overall system efficiency.

1.5.2 SYSTEM EFFICIENCY OF ILLUMINATION VIA LIGHT EMITTING DIODE LIGHTING FROM COAL-FIRED ELECTRICITY

The previous example is of one of the lowest overall efficiencies of daily energy end-use, and since it has two low efficiency steps (energy converted to electricity, and electricity converted to light) we can target one or both to improve energy utilization via enhanced efficiency. If we simply consider the same overall system but replace the lowest efficiency (illumination from an incandescent light bulb), with a roughly 18% efficient LED (light emitting diode) technology, the cumulative efficiency improves by a significant amount. It would rise from 0.567% efficiency to roughly 5.1% efficiency, as shown in Fig. 1.5.

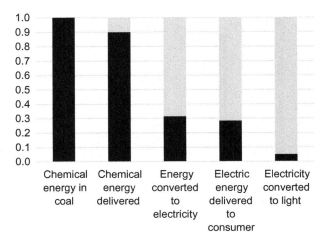

FIGURE 1.5

Efficiency losses by step for coal to LED illumination. *LED*, Light emitting diode.

The much higher efficiency of LED devices makes a very large impact at the final conversion step and is vastly improved over the base case, representing a huge potential energy savings if employed on a large (i.e., national) scale. However, when there is a second major efficiency limiting step, the ultimate efficiency remains low.

1.5.3 SYSTEM EFFICIENCY OF ILLUMINATION VIA LIGHT EMITTING DIODE LIGHTING FROM NATURAL GAS-FIRED ELECTRICITY

Consider next replacing the source with one that has higher efficiencies, especially at the power plant conversion step: natural gas. Using what is known as a combined cycle gas turbine power plant, one can generate electricity at about a 60% efficiency. This is nearly twice the level of conventional coal-fired steam generation. Gas is also less energy intensive to mine and transport, so the ultimate efficiency is roughly twice that of a comparable coal-fired process, and the step-by-step conversion would now look like:

Step 0: Natural gas at the field (no system processes)
   Step efficiency = 1.0; cumulative efficiency = 1.0
Step 1: Extraction and delivery (roughly 5% loss in extraction and delivery)
   Step efficiency = 0.95; cumulative efficiency = 1.0 × 0.95 = 0.95
Step 2: Energy converted to electricity (roughly 40% losses in gas power conversion)
   Step efficiency = 0.6; cumulative efficiency = 0.95 × 0.6 = 0.57

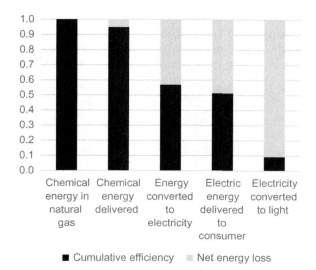

FIGURE 1.6

Efficiency losses by step for natural gas to LED illumination.

Step 3: Transmission of electricity (roughly 10% loss in energy delivered to consumer)
Step efficiency = 0.9; cumulative efficiency = $0.57 \times 0.9 = 0.513$
Step 4: LED lighting (roughly 82% losses in LED lamp conversion)
Step efficiency = 0.18; cumulative efficiency = $0.513 \times 0.18 = 0.09234$

Presenting this information graphically in Fig. 1.6, we clearly see a large improvement in efficiency; however, over the entire process, we still only see a cumulative efficiency of under 10%. This is a relatively low efficiency value within the context of wanting to greatly minimize losses, which puts a spotlight on the thermodynamic penalties associated with chemical conversion via combustion.

1.5.4 SYSTEM EFFICIENCY OF ILLUMINATION VIA LIGHT EMITTING DIODE LIGHTING FROM SOLAR PV TECHNOLOGY

Consider, now, a typical solar PV system running LED lighting. In spite of the common "received wisdom" that solar PV is extremely inefficient and that efficiency is the limiting problem for adopting PV systems on a large scale, the cumulative efficiency is actually not far off from the other dominant fossil fuel technologies. The reason for this is that whereas the initial acquisition step is much less efficient, there are fewer conversion steps involved and, thus, fewer instances of efficiency losses. In the step-by-step conversion, we see the following:

Step 0: Solar energy incident on cell/panel/array (no system process)
Step efficiency = 1.0; cumulative efficiency = 1.0

Step 1: Light converted to electricity and delivered (roughly 85% loss)
    Step efficiency $= 0.15$; cumulative efficiency $= 1.0 \times 0.15 = 0.15$
Step 2: LED lighting (roughly 82% losses in LED lamp conversion)
    Step efficiency $= 0.18$; cumulative efficiency $= 0.15 \times 0.18 = 0.027$

Because of the fewer steps involved, as listed above and as depicted in Fig. 1.7, we calculate a cumulative efficiency of just under 3%, which is competitive (from the point of view of conversion efficiency) with conventional power production using LED lighting.

Looking at Figs. 1.4−1.7, it should be clear that all of these systems (as is generally true for most energy systems) have relatively low overall efficiencies. The more steps there are in the conversion process, the more likely it is that there will be at least one highly problematic step, which limits the total system efficiency. Furthermore, that first inefficient step for solar power is not considered in conventional analyses of stock-based (nonrenewable) energy sources. That is, we do not consider the chemical energy content of coal, oil, or gas that are left unproduced at the mine or at the well. Those unrecovered resources are fundamentally the same as discounting the solar energy not captured and converted by PV cells.

If we assessed the "fuel-to-lamp" efficiencies on an even basis, there would have to be an acquisition step for coal, oil, and gas. For coal and oil, that step would only be about 30% efficient (i.e., there is only a 30% extraction efficiency from the mine or the field), and for gas, the extraction efficiency would be 80% or higher. Thus, the overall efficiency of each system would be reduced by those factors. Alternatively, the 85% "efficiency loss" in failing to convert all of the sunlight striking a solar panel into electricity should be removed from the

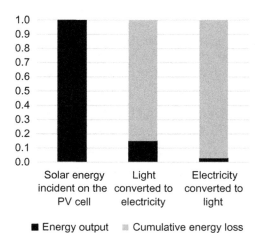

FIGURE 1.7

Efficiency losses by step for solar PV to LED illumination. *PV,* Photovoltaic; *LED,* light emitting diode.

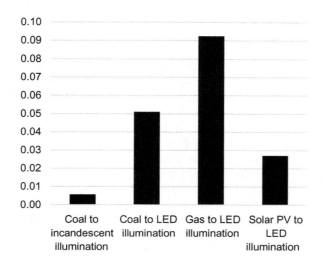

FIGURE 1.8

Net efficiency comparisons for fuel-to-light conversion from coal, gas, and solar PV. *PV,* Photovoltaic.

analysis, just as it is not typically considered for other resources. Indeed, if we treated solar PV on the same basis, the system would be more efficient than any of the other systems, except potentially natural gas, which has the highest resource recovery efficiency of any of the fuels.

Make no mistake, we are not suggesting that solar PV is more competitive than it appears, rather that energy conversion efficiency is not its central problem. The two real challenges that solar PV faces are the limited dispatchability[8] and the front-end capital investment required. The economic inefficiency is largely tied to the low cost of prevailing energy systems, especially in the United States. We simply argue that the focus on the efficiency of capturing light and converting it to electricity is not the defining challenge for solar PV.

1.5.5 FINAL THOUGHTS ON CUMULATIVE EFFICIENCIES

Let us reflect on the four lighting systems discussed in this section using Fig. 1.8. It is evident none of these systems achieve an overall efficiency of even 10%. Lighting systems are not outliers, but rather representative of many energy conversion systems. When we compare the cumulative, net efficiencies, many systems (through their variety of conversion steps) end up with relatively low efficiency performance.

---

[8]Dispatchability refers to the ability to use an energy system when and where needed. Since solar power is intermittent and has low energy density, it has very low dispatchability.

Although a few technologic improvements are available now, the system depicted of coal-fired electricity running an incandescent light was a nearly ubiquitous energy system for more than a century. Low overall efficiency is the norm, not the exception, for energy systems. Low efficiency has never truly prevented the implementation of energy systems. Indeed, even if the energy losses associated with storing electricity in a battery are included, the overall efficiency of the solar PV system remains comparable with the classic coal-fired electricity to power incandescent lights. However, if the inefficiency of initial resource extraction were included for coal or natural gas systems (as is the first step of the solar PV system), the coal and natural gas system overall efficiencies would drop by a factor of three. It is important to keep this in mind when considering efficiencies of other systems.

The reality that system efficiencies are so low, points to the richness of the opportunities in studying energy efficiency. There is far more energy lost than put to effective use. This is central to addressing the environmental impacts of energy use and the limits on access to energy.

1.6 SOURCES OF ENERGY EFFICIENCY INFORMATION

There are several sources that produce information on consumption, production, sources of primary energy, and on environmental data related to energy use and efficiency. These include (1) investor-owned international energy companies, which provide comprehensive energy information; (2) governments, which publish energy data with analysis specific to the purchasing, processing, and overall flow of energy resources; and (3) additional entities, both governmental and nongovernmental, that have been created to report on additional data related to energy use and impacts. Specific sources include

- US Energy Information Administration (EIA Short-term Energy Outlook; Annual Energy Outlook; International Energy Outlook; Energy Explained),
- International Energy Agency (IEA World Energy Outlook; Energy Efficiency Indicators Highlights),
- Organisation for Economic Co-operation and Development (OECD Factbook),
- British Petroleum (BP Statistical Review of World Energy; Energy Outlook),
- Royal Dutch Shell (Energy Scenarios),
- ExxonMobil (Outlook for Energy),
- World Bank (Open Data), and
- World Resources Institute (WRI Maps and Data).

REFERENCES AND FURTHER READING

Books

Blok, K., & Nieuwlaar, E. (2016). *Introduction to energy analysis.* Taylor & Francis.

Culp, A. W., Jr (1991). *Principles of energy conversion.* McGraw-Hill.

Ebenhack, B. W., & Martínez, D. M. (2013). *The path to more sustainable energy systems: How do we get there from here?* Momentum Press.

Fowler, J. M. (1975). *Energy-environment source book.* National Science Teachers Association, Energy-Environment Materials Project, 1742 Connecticut Avenue NW, Washington, DC.

Martínez, D. M., & Ebenhack, B. W. (2015). *Valuing energy for global needs: A systems approach.* Momentum Press.

McLean-Conner, P. (2009). *Energy efficiency: Principles and practices.* PennWell Books.

Weston, K. C. (1992). *Energy conversion.* West Publishing.

Technical Articles and Reports

Armaroli, N., & Balzani, V. (2011). Towards an electricity-powered world. *Energy & Environmental Science, 4*(9), 3193−3222.

Cullen, J. M., Allwood, J. M., & Borgstein, E. H. (2011). Reducing energy demand: What are the practical limits? *Environmental Science & Technology, 45*(4), 1711−1718.

Martínez, D. M., & Ebenhack, B. W. (2008). Understanding the role of energy consumption in human development through the use of saturation phenomena. *Energy Policy, 36*(4), 1430−1435.

Patterson, M. G. (1996). What is energy efficiency? Concepts, indicators and methodological issues. *Energy Policy, 24*(5), 377−390.

Taylor, P. G., d'Ortigue, O. L., Francoeur, M., & Trudeau, N. (2010). Final energy use in IEA countries: The role of energy efficiency. *Energy Policy, 38*(11), 6463−6474.

Energy Information Sources and Reports

British Petroleum (BP). (2017). *BP energy outlook 2017 edition.* Available from <https://www.bp.com/en/global/corporate/energy-economics/energy-outlook.html>.

Organisation for Economic Co-operation and Development. (2016). *OECD Factbook 2015−2016: Economic, Environmental and Social Statistics.* Paris: OECD Publishing. Available from <https://doi.org/10.1787/factbook-2015-en>.

United Nations Development Programme (UNDP). (2018). *Human development reports.* Available from <http://hdr.undp.org/en>.

United States Energy Information Administration (EIA). (2014). *International energy statistics.* Washington, DC: US Department of Energy. Available from <https://www.eia.gov/beta/international/>.

United States Department of Energy Office of Energy Efficiency and Renewable Energy (EERE). (2015). *Energy intensity indicators: Efficiency vs. intensity.* Available from <http://www1.eere.energy.gov/analysis/eii_efficiency_intensity.html>.

World Bank Group. (2014). *World development indicators 2014.* Washington, DC. Available from <http://data.worldbank.org/>.

World Resources Institute (WRI). (2015). *CAIT climate data explorer.* Washington, DC: World Resources Institute. Available from <http://cait.wri.org>.

Other Online Resources

EnergyEducation. ca Encyclopedia. (2018). *Energy intensity.* Calgary, AB. Available from <http://energyeducation.ca/encyclopedia/Energy_intensity>.

Dealing with energy units, measures, and statistics

In this chapter, we review basic energy terms, equivalencies, and statistics essential to understanding energy use and energy efficiency. Key points include:

- common units and prefixes,
- production and consumption definitions, and
- understanding the efficiency ratio.

Energy Efficiency. DOI: https://doi.org/10.1016/B978-0-12-812111-5.00002-0

2.1 THE REASON FOR SO MANY UNITS

When dealing with energy data, we are often bombarded with all kinds of different units. The most basic reason for this is that we use energy in many different ways, as in for cooking, for heating, for lighting, and for transportation. It is quite common to encounter: therms of natural gas; kilowatt-hours of electricity; liters, gallons, and even kilograms of oil; liters and gallons of propane or gasoline; heating values in British thermal units and joules; tonnes of coal; and cords of wood to name some.

As such, energy professionals (scientists, engineers, technicians, analysts, etc.), as well as energy utilities and vendors, often report energy information using different systems of measurement, and often use a wide-ranging language and vocabulary specific to their immediate profession or trade. Old naming and reporting conventions take many years to change, or never change, thus needing a constant reminder of the equivalencies that are needed to translate between all of this tabulated information.

In addition to having to deal with different unit names and definitions, we must also deal with magnitudes of energy content, such as kilowatt-hours or megajoules. Base energy units are very small numbers (a function of how they were originally measured), so to be useful for everyday needs, we must multiply those units sometimes by several orders of magnitude, typically using prefixes and power of 10 notation, as shown in Table 2.1.

Energy use and energy efficiency cannot be properly evaluated without having clarity on the wide range of terms used to quantify energetic sources, fuels, and carriers. For instance, energy is defined as the ability to do work; thus, work and energy share the same units. Indeed, a wide variety of units are used to describe energy itself and an even broader array can be used to characterize the energy content of fuel—or to place energy units as equivalent to certain fuel types. One of the other critical points to be clear about is whether energy is being evaluated or whether power is (the change in energy with respect to time).

It is also worth being clear on the terms "units" and "dimensions," although, sometimes used interchangeably, dimensions are universal combinations of length, time, mass, temperature, light, and number of atoms or molecules. The

Table 2.1 Prefixes and Notations of Various Number Magnitudes

Prefix	Long-Form Notation	Power of 10 Notation
Kilo-	1000	10^3
Mega-	1000,000	10^6
Giga-	1000,000,000	10^9
Tera-	1000,000,000,000	10^{12}
Peta-	1000,000,000,000,000	10^{15}
Exa-	1000,000,000,000,000,000	10^{18}

basic units that we use have specific quantities associated with them, such as meters or feet as "units" of the dimension "length." This would be a base unit, but many units we use combine multiple dimensions. "Velocity" is the ratio of unit "length" (often expressed as "distance" in this application) to unit "time." It is often more convenient to combine all of the dimensions involved in a common phenomenon into a single "unit of convenience."

For example, viscosity, the resistance of a fluid to flow, is commonly reported in "centipoise" but even those of us who use viscosities regularly may need to look up that the dimensions of this unit of convenience are mass multiplied by length multiplied by time. The centipoise quantifies an observable property of a fluid more descriptively than to refer to how many g·cm·s a fluid has. Nevertheless, when ensuring that an equation yields the correct units, it can be very helpful to return to the basic dimensional analyses, to be certain that all of the extraneous dimensions in the equation cancel.

2.2 SYSTEMS OF MEASUREMENT, COMMON UNITS, AND PREFIXES OF ENERGY

There are two systems of measurement commonly used by the world: The United States uses the United States Customary System (USCS) of feet, pounds, and seconds; and the rest of the world uses the International System (SI) of meters, kilograms, and seconds. So, whereas most countries regularly use kilometers, kilograms, liters, and joules in everyday life, the United States still uses miles, pounds, gallons, and British thermal units (Btu). Also, many engineering disciplines continue the tradition of this old, originally British system, so it does not seem like there will be a move away from the USCS any time soon.

Though tedious, converting between various units is fairly straight forward, as energy equivalencies are calculated relative to a certain proportionality. If the conversion is merely needed to convert between two units of energy found in the different measurement systems, such as in converting J to Btu (or vice versa), the proportionality constant will be a ratio of those two values and have units of Btu/J (or vice versa). If the conversion is needed to convert between a certain quantity of fuel and its energy content, then the proportionality constant would also be a ratio of those two values but have units of J/kg of coal or Btu/gal of gasoline (and vice versa).

Let us review some of the important units that we commonly deal with when trying to understand energy use and efficiency (such as the joule and Btu mentioned above).

2.2.1 THE JOULE

The joule (J) is named after the scientist James Joule, who demonstrated via thoughtfully designed experiments involving weights, pullies, paddle wheels, and thermometers that mechanical energy and heat are equivalent units. It is the work done as energy is transferred to an object with a force of 1 N, acting across a distance of 1 m or 1 Nm. Therefore, 1 J = 1 Nm. The newton-meter is also equivalent to 1 kilogram times meter squared divided by a second squared ($kg\,m^2/s^2$). Thanks to the careful experimentation of Sir James, work and energy are now known to be interchangeable concepts, as energy simply is a measurable quantity of work. The more energy something has, the more work it can potentially do.

The joule is the most basic unit of energy in the SI and it is a very small value. For context, let us look at the following scales of energy content and/or energy use:

- juice from one lemon wedge ~ 1000 J (10^3 J or 1 kJ),
- one teaspoon of sugar $\sim 10,000$ J (10^4 J or 10 kJ),
- one laptop battery capacity $\sim 100,000$ J (10^5 J or 100 kJ),
- one 20-oz (591 mL) bottle of cola $\sim 1,000,000$ J (10^6 J or 1 MJ),
- annual energy use of a toaster $\sim 10,000,000$ (10^7 J or 10 MJ), and
- annual energy use of a coffee maker $\sim 100,000,000$ J (10^8 J or 100 MJ).

We can see that by the time we get to energy values beyond using a coffee maker, writing the long form of 100 million becomes tedious, thus the need for prefixes and power of 10 notation.

2.2.2 THE CALORIE

The calorie (cal) is an old unit used for measuring the quantity of heat and is still used in two ways: (1) the calorie, also known as the small calorie, is the amount of thermal energy needed to raise 1 g of water by 1°C at 1 atm of pressure and (2) the Calorie (capital "C"), also known as the big calorie or kilocalorie, is 1000 times the calorie unit, equivalent to the thermal energy needed to raise 1 kg of water by 1°C. The chemical energy content of food that is metabolically available is typically measured in Calories.

To convert from calories to joules, the proportionality constant is roughly 4.2 J/cal or 1 cal = 4.2 J. The big calorie, or kilocalorie, is Cal = 1000 cal = 1 kcal = 4200 J = 4.2 kJ. A typical, healthy adult requires 2000 kcal or 8400 kJ to maintain their weight.

2.2.3 THE BRITISH THERMAL UNIT

The British thermal unit, Btu, is another unit of heat used in the United States. One Btu is the amount of heat required to raise the temperature of 1 lb of water by 1°F. Domestic and commercial heating and cooling systems such as water heaters, boilers, furnaces, air conditioners, and heat pumps in the United States tend to be specified in thousands or tens of thousands Btu/h, which measures the rate at which heat (or cooling) can be transferred (or extracted) by these devices within living and working spaces (*note:* Btu/h is actually a unit of power, which

Table 2.2 Energy Content of Common Fuels

Fuel Type	Heat Value (Btu/lb)	Heat Value (Btu/vol)	Joule/Volume Equivalent
Dry wood	6621	6621 Btu/ft^3	247 MJ/m^3
Natural gas	20,262	983 Btu/ft^3	36 MJ/m^3
Liquid petroleum gas	19,561	87,664 Btu/gal	24.4 MJ/L
Ethanol	11,479	75,583 Btu/gal	21.0 MJ/L
Gasoline	18,659	114,761 Btu/gal	32.0 MJ/L

The Btu is a tiny unit of energy equal to burning one kitchen match, but even so it is equal to 1055 J or 252 cal.
Table created using data presented in The Engineering Toolbox (2016), with adaptation.

we will discuss later on). Also, the energy content of fuels, also referred to as heating values of fuels, are often reported in the United States in terms of Btu per weight or volume, such as in Table 2.2 (we will also speak more on heating value, later).

2.2.4 THE THERM

Related to the Btu is the therm, another common unit of measure used in the United States for commercial and residential natural gas sales. Natural gas companies typically measure their fuel sales by volume and as it turns out, 100 ft^3 of natural gas is roughly equal to 100,000 Btu (10^5 Btu), or 1 therm. Thus, 1 therm = 100 ft^3 of gas. To add to the confusion, American gas suppliers and distributors use Latin abbreviations for 100 (c) and 1000 (m). However, to avoid extreme confusion with SI prefixes, it is common to see on a gas bill listing 1 ccf = 100 ft^3 or 1 mcf = 1000 ft^3 (lower case "C" and "M" Latin abbreviations). In the SI, gas quantities are reported in cubic meters (1 m^3 ~ 35 ft^3).

2.2.5 THE QUAD

The joule, the calorie, and the Btu are very small energy units, developed by early researchers for laboratory-scale measurements. As we saw with the joule and the therm, practical units to describe consumption are much larger, and usually are based on how fuels are sold. Also, as we alluded to in Table 2.1, one of the simplest ways to scale up units is simply to add a prefix indicating additional orders of magnitude as in the summary here:

- Kilo—is one thousand (10^3)—as in kJ. In reference to Btu or cubic feet of gas, it is often shown as "M" or "m" (the Roman numeral for 1000).
- Mega—is one million (10^6)—as in MJ. In using Btu's or cubic feet of gas, it is often shown as "MM" or "mm" (which would be the Roman numeral for a million: a thousand thousand).
- Giga—is one billion (10^9)—as in GJ (1000 million).

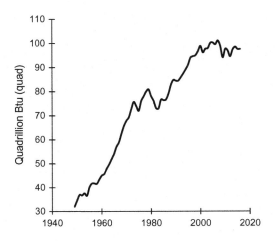

FIGURE 2.1

Primary energy consumption for the United States from 1949 to 2016.
1 quad = 10^{15} Btu = 1.055×10^{18} J or 1.055 EJ.

Figure created using data presented in the US EIA Monthly Energy Review (2018).

- Tera—is one trillion (10^{12})—as in TJ (or 1 million million).
- Peta—is one quadrillion (10^{15})—as in PJ (1000 million million).
- Exa—is one quintillion (10^{18})—as in EJ (1 million million million).

The quad, or Quad, is a unit of measure commonly used by the US Energy Information Administration (EIA) to report on national energy consumption in units of quadrillion Btu. In 2015, for example, the US consumed 97.2 quad or 97.2 quadrillion Btu of primary energy compared to roughly 575 quad for the entire world (including the United States). Recall that 1 Btu = 1055 J, so 97.2 quad is about 102 EJ. Fig. 2.1 displays the most recent historical annual primary energy consumption in the United States, demonstrating the mostly steady increase in consumption with time. There appears to be a leveling off of consumption in recent times, in part to decreased energy intensity of the economy. This is likely related to a combination of financial instability and energy efficiency improvements since 2009.

2.2.6 THE BARREL OF OIL EQUIVALENT

One barrel of oil equivalent (BOE) is another commonly used measure of energy defined as the amount of energy released by burning 1 bbl (42 gal or 160 L) of crude oil. The BOE is used internationally, primarily by oil and gas companies as a means to combine reserves and production of oil and natural gas liquids into a single measure for reporting. The energy equivalent of a barrel of oil depends on

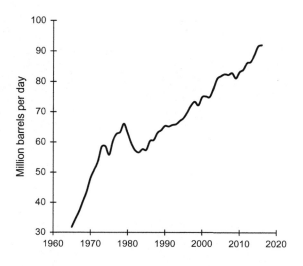

FIGURE 2.2

Historical annual daily global oil production in units of million barrels per day 1965–2016. One barrel = 42 US gallons = 160 L.

Figure created using data presented in British Petroleum's Statistical Review of World Energy (2017).

atmospheric conditions, which results in a range of heating values. Thus, 1 BOE averages 5.8 million Btu (MMBtu or mmBtu, using the latin thousand times thousand), which is roughly equal to 6.1 GJ.

Due to its dominance in the transportation sector, oil is one of the most widely traded energy commodities in the world today, and current global production of oil is quickly approaching 95 million barrels daily. Fig. 2.2 displays the most recent historical annual daily global oil production, demonstrating the mostly steady increase in production with time.

For those who may be curious, a barrel is typically abbreviated "bbl." Early crude oil production was sold in wooden barrels that were painted blue to ensure that the product was not confused with other things (like molasses). So, the abbreviation actually stands for "blue barrels."

2.2.7 THE TONNE OF OIL EQUIVALENT

One tonne (metric ton, 1 tonne = 1000 kg) of oil equivalent (TOE) is a unit of energy defined as the amount of energy released by burning 1 tonne of crude oil. The TOE is a widely used unit in international statistics. The original specification for a TOE was to set 1 TOE to equal 10^7 kcal. To convert TOE to joules, you use a proportionality constant of 42×10^9 J/TOE. That is, 1 TOE = 42×10^9 J or 42 GJ. Each tonne of oil has a slightly different composition and density, so this value does vary slightly from one tonne of oil to the next tonne of oil.

The TOE is used for large amounts of energy and it is quite common to find units of MTOE (1 million TOE) and units of GTOE (1 billion TOE). A smaller unit of energy also related to the TOE is the kilogram of oil equivalent (kgoe). 1 kgoe = TOE/1000. Also, conversion between TOE and BOE depends on the density of oil, which varies by location (1 TOE ∼ 6.8−7.5 BOE).

It is quite common in international reporting to display global energy consumption from all energy sources (i.e., oil, natural gas, coal, nuclear, hydro, and "new" renewables) on a MTOE basis. Fig. 2.3 displays the most recent historical

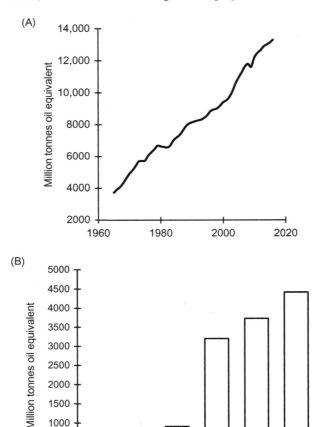

FIGURE 2.3

Historical annual global energy consumption on a MTOE basis from 1965 to 2016 (A) as well as the distribution of energy consumed by resource for the year 2016, also on a MTOE basis (B). 1 MTOE = 10^6 TOE = 42×10^{15} J = 42 PJ.

Figures created using data presented in British Petroleum's Statistical Review of World Energy (2017).

Table 2.3 Energetic Equivalencies to 1 kg Coal in SI Units

0.34 kg coal equivalent	1 kg crude lignite
0.56 kg coal equivalent	1 kg fire peat
0.57 kg coal equivalent	1 kg firewood
0.60 kg coal equivalent	1 m^3 town gas
0.72 kg coal equivalent	1 kg lignite briquette
0.97 kg coal equivalent	1 kg hard coal coke
1.00 kg coal equivalent	1 kg hard coal
1.14 kg coal equivalent	1 kg anthracite
1.35 kg coal equivalent	1 m^3 natural gas
1.52 kg coal equivalent	1 kg fuel oil
1.59 kg coal equivalent	1 kg gasoline

Source: The European Nuclear Society Encyclopedia (2017).

annual global energy consumption on a MTOE basis (A) as well as the distribution of energy consumed by resource for the year 2016, also on a MTOE basis (B). Despite the inroads made by renewables in recent years, oil, gas, and coal continue to dominate global energy consumption.

2.2.8 MASSES OF COAL EQUIVALENT

One tonne of coal equivalent (TCE) is a unit of energy defined as the amount of energy released by burning 1 tonne (i.e., metric ton) of coal. The TCE is another widely used unit in international reporting, since coal remains a historically important fuel for smelting, electricity generation, and for space heating in countries such as China. Indeed, China has traditionally converted all of its energy statistics into TCE base units. Coal has various grades; thus, a TCE also varies by grade. Based on high heating and low heating values, 1 TCE ranges from roughly 29.31 to 31.52 GJ.

A smaller unit of energy related to the TCE is the kilogram of coal equivalent (kgce), which is specified as 7000 kcal ($\sim$29.3 MJ). It is the reference unit commonly used for comparing the energies of various other energy carriers as shown in Table 2.3.

2.2.9 ENERGY CONTENT OR HEATING VALUE OF END-USE FUELS

The transformation of chemical energy to heat is one of the world's most common energy end-use conversions and remains the basis of most of the world's electric power and transportation sectors, as well as the basis of much of the world's residential and commercial heating systems. The energy content of the end-use fuels used for these conversions is typically characterized by the heating value of the fuel per unit of mass. Heating value may be reported either as lower heating value (LHV), also known as net calorific value, or as higher heating value (HHV), also known as gross calorific value. The technical definition for LHV is "the amount of heat released by combusting a specified quantity (initially at 25°C) and returning the temperature of the combustion products to 150°C." Since the temperature is above the boiling point of water, it does not consider that the latent heat of vaporization is recovered in the process. On the other hand, HHV is defined as "the amount of heat released by a specified quantity (initially at 25°C) once it is combusted and the products have returned to a temperature of 25°C." As it returns to a comfortable ambient temperature, this value does consider capturing the latent heat of vaporization.

Based on the above definitions, HHV represents the total amount of heat that can be generated via combustion. Assuming the combustion exhaust gases can be cooled sufficiently to recover the water produced during the combustion process, one can utilize most of the heat content of the original fuel. LHV assumes that recovery of the exhausted water vapor is not possible. Thus, the ratio of HHV to LHV represents the efficiency of fuel utilization based on the ability to recover the water generated during the combustion reaction, because exhausted water vapor carries with it unused energy. Table 2.4 displays high, low, and high/low ratios of heating values of various fuels using the SI. Note that energy content for coals and biomass fuels can be assessed on a wet basis or dry basis. Since grades of coal and biomass can contain significant amounts of water when mined or harvested, a wet-basis LHV would have to account for the energy required to evaporate water, which is 2.26 MJ/kg at standard conditions.

Table 2.4 High, Low, and High/Low Ratio of Heating Values of Various Fuels

Fuel	HHV (MJ/kg)	LHV (MJ/kg)	HHV/LHV
Hydrogen	141.88	119.96	1.18
Propane	50.22	46.28	1.09
Natural gas and LPG	50.14	46.60	1.08
Gasoline (conventional)	46.52	43.44	1.07
Diesel (low sulfur)	45.56	42.60	1.07
Crude oil	45.53	42.68	1.07
Coking coal (wet basis)	29.86	28.60	1.04
Ethanol	29.84	26.95	1.11
Bituminous coal (wet basis)	27.26	26.12	1.04
Coal (wet basis)	23.96	22.73	1.05
Farmed trees (dry basis)	20.58	19.55	1.05
Forest residue (dry basis)	16.47	15.40	1.07

Table created using data presented in the Pacific Northwest National Laboratory's HyARC Calculator Tools (2018).

2.2.10 THE WATT-HOUR, KILOWATT-HOUR, AND THE TERAWATT-HOUR

A watt-hour (Wh) is a unit of convenience of electrical energy, which has been adopted widely. Many international energy statistics are provided in watt-hour units or multiples thereof. The reason why we say that the watt-hour is a unit of convenience is that electricity is, in fact, a unit of power, which is equal to energy divided by time, and has units of J/s. So, a watt-hour is technically energy divided by time multiplied by time again, which is redundant for a base unit.

The watt-hour is therefore an energy equivalent unit of convenience, and as such, the watt-hour is meant to convey the consuming or expending of 1 W of power for a duration of 1 hour. Since a watt is equal to a joule per second, and a joule is an extraordinarily small unit of energy, a watt-hour is also quite small. Thus, the kilowatt-hour (kWh, 1000 Wh) is more commonly used. Indeed, the kWh is the base unit used for pricing electricity consumption at the residential and commercial levels. We often read our utility bills priced in cents per kWh.

To convert kWh to joules, you would multiply kWh by 3.6×10^6 J. Thus, 1 kWh = 3.6 MJ. Because we are talking about the energy transformation from a fuel to electricity, this conversion is equivalent to saying that 3.6 MJ is the amount of energy converted if work is done at the rate of 1000 W consumed or expended for 1 hour. This does not necessarily mean that it takes 3.6 MJ of fuel to generate 1 kWh; however, as we saw in Chapter 1, Introductory Concepts, fuel conversion efficiency is roughly 33% for average thermal electric power processes. That means, it would actually require 10.8 MJ to generate 1 kWh of electrical energy from traditional thermal power plants. Thus, the 1 kWh to 3.6 MJ conversion assumes 100% fuel conversion efficiency. (We will look at this again from a different perspective in Section 2.3.1.2.)

We want to emphasize that one of the most common sources of confusion in energy discussions is around watts and watt-hours. As we have explained, a watt is a unit of power rather than a unit of energy, defined as 1 J divided by second. Since the joule is a unit of energy, watt is a unit of energy per unit time. It is quite common to see an incorrect usage of "watts per hour," which would be a unit equivalent to energy divided by time divided by time, instead of the proper watt-hour, which is simply energy.

Finally, a larger electrical energy unit used in energy statistics reporting is the terawatt-hour (TWh). Recall that the prefix "tera" is equal to 10^{12}, so 1 TWh = 10^9 kWh. Country and world statistics are reported in TWh. Fig. 2.4 displays the most recent historical annual global electricity generation on a TWh basis. It is remarkable that electricity generation in 2016 was 2.5 times larger than it was in 1985, speaking to the incredible demand for electricity worldwide, particularly in rapidly expanding economies such as in China and India. With over a billion people still without reliable access to electricity worldwide, it is expected that growth in electricity generation will continue into the foreseeable future. This also speaks to the likely need for different, more efficient generation

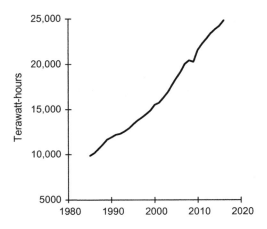

FIGURE 2.4

Historical annual global electricity generation in units of terawatt-hours (TWh) from 1985 to 2016. 1 TWh = 10^9 kWh.

Figure created using data presented in British Petroleum's Statistical Review of World Energy (2017).

technologies, because with current technology, three units of thermal energy (derived from fuel) are required to generate one unit of electrical energy. Increasing electricity generation to serve 1 billion additional people would mean a massive increase in total primary energy demand, using current generation technologies and low efficiency thermal processes.

2.2.11 VISUALIZING ENERGY UNITS AND THEIR EQUIVALENCIES

Figs. 2.5–2.7 provide an additional perspective of the multitude, and magnitude, of differences between equivalent energy units. The majority of these units originated as small amounts of energy useful for confirming laboratory-scale experiments most related to equivalencies and heating values. These units also suffice to describe market prices of energy ($/bbl or cents/kWh), but in order to describe global, or even regional or city-wide energy consumption, we must use billions, trillions, or even quadrillions of these units.

Some of the energy units used in Figs. 2.5–2.7 and their equivalencies are listed below:

- 1 Btu is the equivalent of: 1055 J; 252 cal (note "calories" with a small "c"); 0.203Wh.
- 1 MCF of gas has an approximate energy content of 1 million Btu, the equivalent of 10 "therms" (a term often used in consumer gas bills).

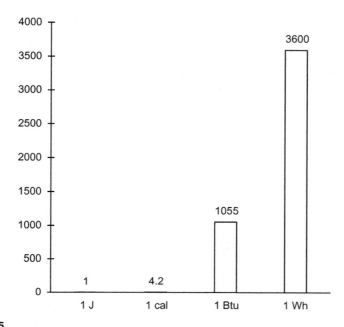

FIGURE 2.5

Visual representation of unit equivalencies of a joule (J), calorie (cal), British thermal unit (Btu), and a watt-hour (Wh). The bar chart shows energy units equivalent in joules (1 cal = 4.2 J, 1 Btu = 1055 J, etc.)

- 1 BOE represents the typical energy content of a barrel of oil: 5.8 million Btu.
- 1 TCE represents the typical energy content of a metric tonne of mid-rank coal: 27.8 million Btu.
- 1 TOE represents the energy content of a metric tonne of a common crude oil: 39.7 million Btu.

Notice also in Figs. 2.5–2.7 that the scale changes needed to compare the variety of energy equivalent units. Whereas the watt-hour dwarfs the joule, the watt-hour is dwarfed by the cubic foot of gas, which itself is dwarfed by the BOE, the TCE, and the tonne of oil equivalent.

2.3 POWER AND ITS EQUIVALENCIES

As mentioned earlier, power is the change in energy with respect to time, or energy flow per unit time. It refers to the rate at which energy is converted to work, such as in energy conversion, in energy production, and in energy through-put. The most commonly used unit for power is the watt, or multiples of it. One

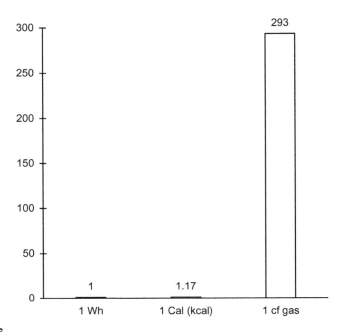

FIGURE 2.6

Visual representation of unit equivalencies of a watt-hour (Wh), kilocalorie (kcal), and a cubic foot (cf) of gas. The bar chart shows energy units equivalent in Wh (1 Cal = 1.17 Wh, 1 cf = 293 Wh, etc.)

watt is defined as 1 J of energy per second, based on the relationship between power, P, energy, E, and time, t:

$$P = \frac{E}{t}$$

Watts are relatively small units; thus, they are appropriate for describing the power usage of small devices and appliances. Household and commercial devices, appliances, and equipment are tested by the manufacturer and labeled with power ratings based on constant inputs. Standard general use lightbulbs (i.e., ambient lighting), for example, have power ratings on the order of 10 W (LED lamps) to 100 W (incandescent lamps). Space heaters, microwave ovens, and refrigerators are likely to be rated at more than 1000 W, or 1 kW. If the input power ratings for these appliances are multiplied by the time of their use in hours (i.e., how long they have been "on and running"), then the resulting value is in kWh, the energy equivalent unit of convenience described in Section 2.2.10.

Power output or generation, referring primarily to the power output of electricity generating power plants, is also rated in watts. A large household solar photovoltaic array may have an output of 5 kW, whereas utility-scale power plants are

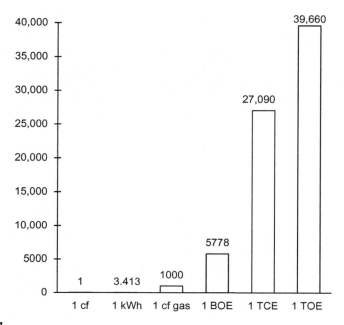

FIGURE 2.7

Visual representation of unit equivalencies of a cubic foot (cf) of gas, a kilowatt-hour (kWh), a thousand cubic feet (mcf) of gas, a barrel of oil equivalent (BOE), a tonne of coal equivalent (TCE), and a tonne of oil equivalent (TOE). The bar chart shows energy units equivalent in cf (1 kWh = 3.413 cf, etc.)

more likely to have outputs of millions of watts, or megawatts (MW). Individual large-scale wind turbines, towering some 100 meters, are in the range of several MW. Large-scale power plants (coal, natural gas, and hydropower) are likely to produce ~10−100 times as much: on the order of hundreds of MW or even 1000 MW, known as a gigawatt (GW).

Common alternative, but equivalent usages and units of power are described below.

2.3.1 LOAD AND CAPACITY

A unit of power can be created from any unit of energy combined with any unit of time. We can measure power in joules or calories per second, Btu per hour, kilojoules per month, and so on. There are unlimited possible combinations and most exist for the convenience of the user. One of the most prominent uses is with respect to energy load or energy capacity, which looks at the amount of energy passing through a physical system per unit time.

2.3.1.1 Building loads and equipment capacities

Good examples that involve the measuring or estimating of energy loads and equipment capacities can be found in residential and commercial building energy. With respect to building energy and thermal "loads," we often calculate or measure the heat energy that needs to be added to or removed from a room or building to maintain a steady state thermal balance (as well as to control moisture for human comfort). Thermal loads, typically measured in units of Btu/h in the United States, include internal building loads from the heat generated by the occupants, by equipment, and by lighting, as well as include external building loads from the heat generated by the sun, by the air, and by latent heat from moisture in the air.

Likewise, heating and cooling systems that are used to regulate or control heating and cooling loads can be passive, using thoughtful building design, or they can be active, using equipment that utilize gas or electricity, as in room air conditioners, and "HVAC" and hot water systems to provide process heat to the space. The amount of heating or cooling that a piece of equipment can provide is known as its "capacity." As an example, Table 2.5 provides an ENERGY STAR room air conditioner sizing chart, relating room area to cooling capacity needed.

Additional loads that are considered in building energy include home appliance and office equipment loads, also known as plug loads, as well as lighting loads, which are both measured in watts. These electrical loads are also referred to and measured by the power that they draw from the electricity that is supplied and delivered by the electric power grid.

Finally, insulating materials can be used to reduce thermal loads through spaces and building materials. It is common to see "U values" and "R values," which both represent heat flow or resistance to heat flow through a material (in units of W or Btu/h).

Table 2.5 Cooling Capacity Sizing Chart for a Residential Room Air Conditioner up to 1000 ft^2. 1 m^2 = 10.76 ft^2. Sizing is Affected by Solar Gain, Occupancy, and Other Factors

Area to Be Cooled (ft^2)	Capacity Needed (Btu/h)
100−150	5000
150−250	6000
250−300	7000
300−350	8000
350−400	9000
400−450	10,000
450−550	12,000
550−700	14,000
700−1000	18,000

Table created using data presented in ENERGY STAR's Properly Sized Room Air Conditioners (2018).

2.3.1.2 Electricity generation and capacity

Electric power plants are designed to convert the energy found in fuels like coal, gas, uranium, and biomass, as well as kinetic and light energy from the wind and sun, into useful electricity. In order to "keep on the lights," these energy conversions occur continuously, thus generating power (i.e., energy over time). As such, electric power is the rate of energy flow or consumption of fuels used to produce electricity, and power plants are rated by their capacity to produce and deliver electric power.

For example, consider a coal-fired power plant rated at 1 GW of electrical output. This rating is the electrical output capacity of the power plant, and a 1 GW power plant has the capacity to put electric power into the grid at a rate 100 times higher than a 10 MW plant. The input load consumed by the plant in order to generate 1 GW of electricity is referred to as the plant's "heat rate," or in this case, the rate of thermal energy supplied by the coal fuel. Recall from Section 2.2.8 that the energy content of coal is roughly 30 GJ/t, so if the power plant is operating at 33% efficiency, the coal thermal energy input needed to output 1 GW of electrical power would be

$$\text{Input} = \text{Output}/33\% = 1\ GW/0.333 = 3\ GW = 3 \cdot 10^9\ J/s \times 3600\ s/h = 10,800\ GJ/h$$

Using the 1 TCE $\sim$ 30 GJ conversion, this is equal to needing 360 tonnes of coal per hour to operate the plant. If operating at full capacity (i.e., 24 h/day), this plant would require 360 tonnes/h $\times$ 8760 h/year = 3.15 million tonnes of coal per year.

For comparison, Fig. 2.8 displays daily fuel input requirements for a 1 GW power plant with and without plant efficiency losses for various fuels. So, the taller the bar, the more input is required. Bear in mind that the practical units shown for each resource are not comparable to each other, but rather Fig. 2.8 is presented to provide a general sense of the amount of fuel needed using their fuel-specific reporting metric (i.e., by volume or by mass). A thousand standard cubic feet (MSCF) of gas[1] has a little over 1/50th the mass of a tonne of coal. So, even though seven times more MSCF are required than tonnes of coal, the gas-fired power plant is still using one-sixth as much fuel. The point of Fig. 2.8 is to show a different perspective of the significance of the fuel requirements and the efficiency losses in each power plant.

It is also important to understand that the 1 GW output of each power plant above is its stated "nameplate" capacity, which represents the approximate maximum rate of electric power generation the plant can achieve. The actual production is reduced by the "load factor" or "capacity factor." Load factor is defined as

---

[1]For practical purposes, at surface conditions, where gas is sold and consumed, MCF always implies MSCF. For petroleum engineers, evaluating reservoir behavior at higher temperatures and much higher pressures, there can be a big difference and it becomes important to know whether the gas is being described at reservoir conditions or standard. For those living and working on the surface, it is generally fair to assume people are referring to MSCF.

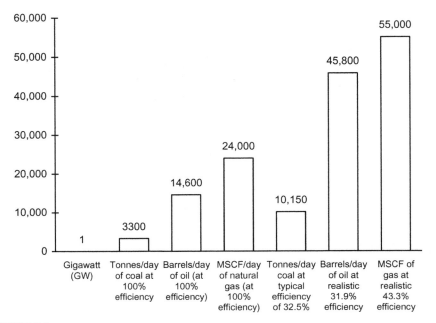

FIGURE 2.8

Daily fuel requirements for a 1 GW power plant with and without efficiency losses for coal, oil, and natural gas. *Note:* not presented using a uniform mass and/or volume basis.

"the ratio of the number of hours of electricity generated in a year to the total number of hours in a year." Capacity factor is "the ratio of the number of units of electricity generated to the number of units the plant can generate if in 100% operation, over the course of a year." For coal-fired and nuclear power plants, the load and capacity factors can be quite high, on the order of 90%.

For utility-scale solar and wind power plants, average capacity factors are closer to 25% and 35%, respectively, so it is especially important to be careful not to confuse nationally and internationally reported installed capacity with actual power generation. For example, Fig. 2.9 shows the 2016 installed utility-scale solar and wind capacity for the world, separated by region. To estimate actual electricity production, these values should be multiplied by 0.25 and 0.35, respectively. (Capacity factors for renewable energy obviously can vary substantially from these average values based on region.)

2.3.1.3 Large-scale energy consumption and loads

There is one last important example of energy load. As we saw in Fig. 2.1, it is possible to talk about large-scale energy consumption on a quadrillion Btu or exajoule per year basis (quad/year or EJ/year). This is a common means to quantify the annual rate of societal energy use, which is merely another unit of power.

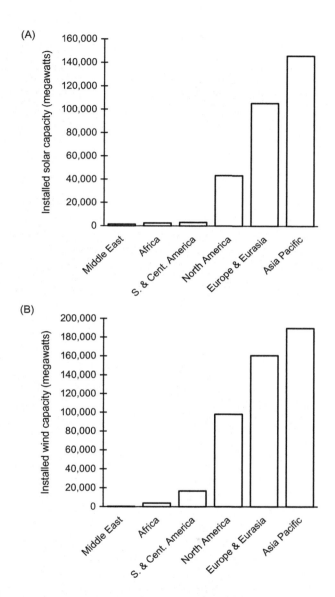

FIGURE 2.9

2016 global installed utility-scale solar (A) and wind (B) capacity by geographic region, in MW.

Figures created using data presented in British Petroleum's Statistical Review of World Energy (2017).

Using the language for building energy flows and power generation, you can equivalently call these statistics "country energy loads." It is also common to see global annual statistics reported in kWh/year or TWh/year; however, this convention is rather redundant, since it would be just as valid to report this information in kW or TW. For example, in Fig. 2.4, we see 2016 global electricity consumption reported as 24,816 TWh/year. If we just divide this value by 8760 h/year, we would find that the world consumed at a rate of 2.83 TW in 2016, which seems to be a much simpler metric to report.

2.3.2 **FLUX**

Another useful power equivalence, flux, deals with energy flow through a unit area, and is measured in watts per square meter (W/m^2). In the study of visible light, luminous flux is measured in lumens (lm), and luminous emittance or illuminance measured in lux (lm/m^2). In the study of radiant energy, particularly in solar energy analysis, power is commonly measured as a radiant flux, emittance, irradiance, or insolation. In the study of heat transfer, thermal flux is a heat rate per unit area, which can be equivalently measured as $Btu/s/ft^2$. We will talk briefly about the first two below.

2.3.2.1 *Light and illuminance*

Light is a category of electromagnetic radiation, which has been found to have both the characteristics of a wave and of particle. The "intensity" of light describes the total energy conveyed. The term "brightness" describes the light intensity as perceptible to the human eye. There are three basic ways to describe light intensity:

- light emitted at a source (radiance),
- how bright is the light at a specified distance, and
- how brightly does the light illuminate a given area.

Light is useful work in enabling people to see in otherwise dark environments. The light must illuminate the desired subject, while being reasonably comfortable to the viewer. The color rendition and steady brightness of light are all important to its comfortable use, as well the focus and intensity of it.

Furthermore, prompt illumination has become an expectation of modern electric lighting. Older fluorescent lights, though needed to warm up. Therefore, in cold weather, their effectiveness was impaired by either the need to turn them on before the light was truly needed, or deal with waiting for the light to reach desired levels.

In terms of units and measures, illumination is evaluated in two ways: (1) the total amount of light at the source, radiance and (2) the intensity of light on a surface, illuminance. As depicted in Fig. 2.10, radiance is measured in candelas and the intensity of one candela located 1 ft away from the source is 1 foot-candle (ft-cd), while the illumination onto a 1 ft^2 area is 1 lm. In terms of meters, the intensity of 1 cd is 1 lm and the illumination on 1 m^2 is 1 lx. Lumens are the unit of lighting measure most closely associated with efficiency of residential lightbulbs or lamps.

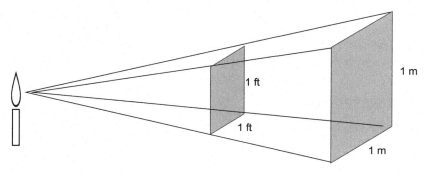

FIGURE 2.10

Depiction of the relationship between candle intensity and illumination onto a unit area.

Table 2.6 Average Annual Solar Flux and Insolation Values for Various Cities.

City	Solar Fux (W/m²)	Insolation (kWh/m²/day)
Algiers	204	4.9
Boston	174	4.2
Perth	230	5.5
Beijing	169	4.1
London	128	3.1
Berlin	104	2.5
Sao Paolo	198	4.8
Riyadh	251	6.0

Table created using data presented in NASA's Prediction of Worldwide Energy Resources (2016).

2.3.2.2 Solar flux

Our sun is continuously converting massive amounts of matter into energy that leaves in the form of electromagnetic radiation and travels virtually unobstructed through space to Earth at a flux of roughly 1370 W/m². Earth's geometry, as well as various atmospheric processes reduces this to an average value of about 170 W/m², which represents the average annual solar flux incident on Earth's surface. In other words, over a 1-year period, the 170 W/m² of constant solar energy equates to the sun providing roughly 5.4 GJ of energy onto 1 m² of Earth's surface. This is approximately the energy that can be extracted from 0.2 tonnes of coal, 9/10 of a barrel of oil, or 49 therms (49 ccf $\sim$ 140 m³) of natural gas.

Of course, at specific times of the year, and at specific latitudes, solar "insolation" can vary widely. But, the average values are important especially when considering the use of solar collectors, both for heat and electricity needs, as well as for biomass growth over year long timeframes. These values can be obtained from insolation data collected annually and over several decades. Often average annual values are displayed in both flux units of W/m² and in insolation units kWh/m²/day. Values for a handful of cities based on 2017 measurements are presented in Table 2.6.

In the case of solar energy harvesting, the combination of this insolation information, along with panel efficiency ratings can be used to assess the expected output, adjusted for installation and other features, such as tilt angle, panel tracking ability, and building and landscape shading characteristics at the site.

2.3.3 HORSEPOWER

As the name implies, the horsepower (hp) is a unit of power, or energy per unit time. The term itself was coined by inventor James Watt, based on his observations and reasoning on the rate of work done by horses lifting loads from coal mines. He estimated that a horse, doing sustained work could lift 33,000 ft-lb/min. (That could be 330 lb raised 100 ft/min, for example.) The motivation behind Watt's horsepower term was to compare the power of horses to that of his then state-of-the-art steam engine technology. Using his crude calculations, Watt advertised that one of his engines could produce power sufficient to replace 10 load-lifting horses, or 10 hp.

Horsepower refers to a relatively ideal power figure, especially as engines quickly replaced draft animals for most tasks. It is now measured in various ways related to torque rather than linear force. "Brake horsepower" (also, "shaft horsepower") is the horsepower measured at the output shaft of engines, motors, or turbines. The horsepower of electric motors also can be measured based on electrical input. "Indicated horsepower" is the power produced in reciprocating engines, determined by the measurement of pressure in the piston cylinders. In terms of equivalencies, 1 hp has an electrical equivalent of 746 W and a heat equivalent of 2545 Btu/h. (The "metric horsepower" is defined slightly differently as 4500 kg m/min, and equivalent to 0.986 hp.)

Today, horsepower is probably most closely associated with automobile engine power (although also reported for smaller equipment like lawnmowers and generators). As automobile engines produce torque, a device called a dynamometer can be used to measure equivalent horsepower, placing a load (i.e., a brake) on the engine and measuring the power that the engine can produce against it. In automobile engine testing, torque is measured by varying the engine speed, which itself is measured in revolutions per minute (RPM). Horsepower, then, is calculated by taking torque and multiplying it by RPM and dividing by 5252. Personal automobiles typically have engines ranging from 70 to 250 hp, depending on size and regional preferences.

2.4 PRODUCTION AND CONSUMPTION DEFINITIONS AND STATISTICS

Energy access, availability, and use are reported with a wide range of statistics. In tracking efficiency, it will be important to understand the meanings and uses of the range of statistical information.

2.4.1 RESOURCES AND RESERVES

Reserves of fossil fuels refer to the amount of coal, oil, or gas resources located in subsurface formations (this would also apply to uranium ore). While there are many different reserves classifications, the most commonly used in oil and gas is probably "proved recoverable reserves." This refers to the amount of oil or gas that has been proved to exist by successful production and is believed to be recoverable under existing economic and technologic constraints. This should be a reasonably conservative number, since it requires direct evidence of the resource being present and capable of production. It does not consider possible reserve additions resulting from additional drilling, or increased price, or new technologies. In general, the proved reserves of an oil field are less than one-third of the oil in place (i.e., in the ground).

On the other hand, the term "resource base" is the most optimistic and the least certain data for fossil fuels. It is based on estimations of how much exists, with no regard to how much can ever be produced. It also does not require proof of existence. Many formations that appear from surface measurements to hold great oil or gas potential prove to be barren. Nevertheless, as discoveries are made, new provinces open, revealing oil and gas that may not have been part of the resource base estimates. This explains why the resource base has continuously grown throughout the petroleum era. It is not that new oil and gas are forming as fast as we extract them, it is that there is simply far more of them in the earth than had been expected.

Both reserves and resource base can be defined for a given field, a region, a nation, or the world. Reserves must always be less than the estimated resource base. One of the misuses that appears from time to time is to compare the resource base of one energy source to the reserves of another. For example, during fears of oil shortages, it has been common to compare the resource base of oil shale (i.e., extracting solid organic matter called kerogen from shales and retorting it at high temperature to convert the solid matter to liquid "synthetic crude"). Since this process has never proven to be commercially successful, there can be, by definition, no reserves. On the other hand, gas and oil produced from shales through massive induced fractures in long lateral sections of a wellbore have proven commercial and radically transformed the understanding of global oil and gas resource base—and added enough reserves to depress prices severely.

Fig. 2.11 provides an informative visual for understanding how reserves classifications have been developed by the Society of Petroleum Engineers. The total area of Fig. 2.11 represents the resource base, "total petroleum in place" in this case. The reserves are located at the top, with proved reserves at the far left. To the right, the confidence or probability decreases (from proved to possible). Going down the chart are resources that are subcommercial, under current conditions, and then prospective resources—those which have not been discovered. There are also contingent and prospective unrecoverable resources.

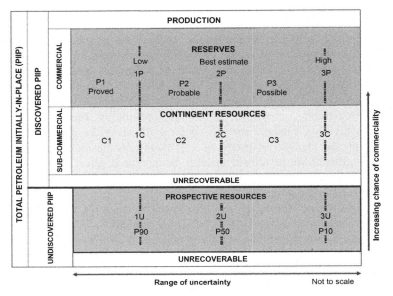

FIGURE 2.11

Petroleum resources and reserves classifications, based on chance of commerciality and physical recoverability.

Figure reproduced with permission from information and imagery presented in the Society of Professional Engineers Petroleum Resource Classification Framework (2018).

Since reserves already refer to recoverable quantities, there would be no unrecoverable fraction in the reserves category. Increased price and technology can move contingent resources into reserves, as drilling discoveries can advance prospective resources to reserves. For oil and gas especially, it is important to note that there cannot be "proved reserves" until at least one successful well has confirmed the commercially producible quantities of oil and gas.

Again, of note is that the subcommercial, undiscovered, and unrecoverable resources are not counted against the efficiencies of petroleum systems. Only the "captured" or "anthropogenically controlled" energy produced to the wellhead is counted as input into the energy system. Of course, it is still desirable to increase the "recovery factor," which equates to moving resources from the contingent or unrecoverable categories into the proved reserve category. Similarly, exploratory successes move resources up from the "prospective" to the "reserves" category. None of this, though, changes how we describe the energy efficiency of the system.

Similar resource and reserves terms can be used to describe coal, and Figs. 2.12–2.14 display the 2016 global oil reserves, gas reserves, and coal reserves by region, using their representative base unit. Though not necessarily indicative of the resource base (though likely), it is obvious that there exists an unequal distribution of primary energy reserves across the planet.

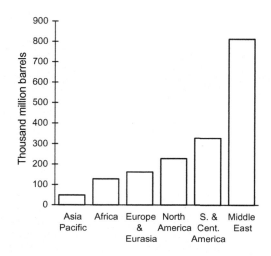

FIGURE 2.12

2016 global oil reserves by geographic region, in thousand million barrels.

Figure created using data presented in British Petroleum's Statistical Review of World Energy (2017).

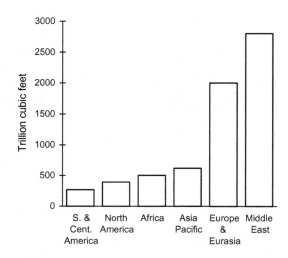

FIGURE 2.13

2016 global gas reserves by geographic region, in trillion cubic feet.

Figure created using data presented in British Petroleum's Statistical Review of World Energy (2017).

It is important to understand that these classifications, except perhaps for biomass, are not as appropriate for the renewable energy resources. Every resource has a resource base: how much of it exists, and there is production for every resource, of course, however, solar and wind, in particular, have no real reserves.

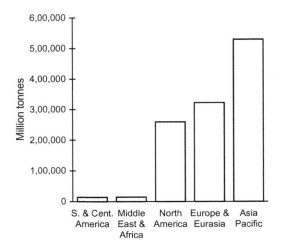

FIGURE 2.14

2016 global coal reserves by geographic region million tonnes.

Figure created using data presented in British Petroleum's Statistical Review of World Energy (2017).

Table 2.7 Plausible Renewable Energy Flows That Could Be Tapped as Anthropogenically Controlled Energy

Plausible Energy Flow	EJ/year	TW
Runoff[a]	64	2
Biomass[a]	390	12
Geothermal[b]	1023	32
Wind, waves, currents[b]	11,700	366
Solar radiation on land[b]	765,000	24,000

[a]*Smil (2008).*
[b]*Boyle (2004).*
Table created using a composite of information presented in Smil, V. (2008) and Boyle, G.B. (2004).

(Hydropower perhaps can claim reserves in applications where substantial amounts of water, and thus potential energy, are stored behind dams.) The closest analogy to reserves would be evaluations of what fraction of the resource base can plausibly be tapped as a flow such as in Table 2.7. It is obvious that there exists a massive potential to produce useful energy from renewable flows; however, the variable and diffuse nature of these resources as well as their lack of portability and storage have thus far prevented them from taking substantial portions of the global energy market.

2.4.2 **TOTAL PRIMARY ENERGY SUPPLY AND TOTAL FINAL CONSUMPTION**

Total primary energy supply (TPES) and total final consumption (TFC) are two international statistics that are commonly found when looking at energy end-use and efficiency data. TPES is defined by the Organisation for Economic Co-operation and Development as "energy production plus energy imports, minus energy exports, minus international bunkers, then plus or minus stock changes." TPES is essentially the total amount of energy that a country has access to at any given time, usually reported on in monthly and yearly intervals.

Energy is used and lost at every step of the energy conversion process, which must be accounted for. Energy users include both the end-use consumers and intermediate conversion processes. Energy analyses categorize the end-users in consumption sectors. They can be enumerated as follows:

- electric power generation,
- transportation,
- industrial,
- commercial, and
- residential.

The electric power sector is not technically an end user, but a "throughput" sector, as its entire role is to generate the useful energy carrier electricity. It is viewed as its own sector simply because of the large amount of energy consumed (and lost). Electric power and other energy conversion sectors lose a substantial amount of energy in the conversion process and must be accounted for with positive and negative throughputs. Once this lost energy is accounted for, what is left is TFC. That is, TFC is equal to TPES minus conversion process inputs plus conversion process outputs. A representation of how TPES and TFC interact is provided in Fig. 2.15.

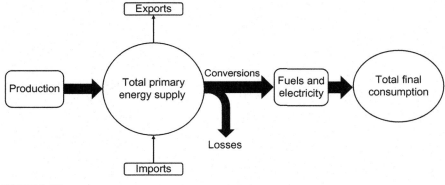

FIGURE 2.15

Schematic representation of the relationship between TPES and TFC. *TPES*, Total primary energy supply; *TFC*, total final consumption.

Figure created using information and imagery presented in Energy Education - Total primary energy supply [online], J.M.K.C. Donev et al. (2017), with modification.

2.4.3 ENERGY INTENSITY

Energy intensity was discussed in Chapter 1, Introductory Concepts; however, it is worth mentioning again that it is commonly reported in annual energy reporting. Every human activity in modern societies requires energy inputs, and the ratio between energy use and economic strength, typically as measured by the gross domestic product (GDP), is referred to as the "energy intensity" of that economy. Indeed, the economic growth of the 20th century was built largely on cheap and abundant fossil fuel energy. There has been some recognition for decades that energy cannot be harvested indefinitely. It has also been clear that environmental impacts make it undesirable to continue high energy intensity.

In recent times, energy consumption has not grown at the same rate as GDP for most countries. The primary reason has been that service sectors, particularly in developed nations, have become the dominant driver of economic growth. Banks, insurance companies, and software companies, for example, can generate more money for less energy and fewer labor costs than a large steel factory can. Also, the integration of more efficient technologies and processes have helped reduce energy demand, thereby generating more money or capital for less energy used.

2.4.4 PER CAPITA ENERGY CONSUMPTION

Per capita energy consumption (PCEC) indicates the energy use per person in a particular region or country. The most widely used metric for reporting PCEC is the oil equivalent, either in kilograms of oil equivalent (kgoe) or tonnes of oil equivalent (TOE). PCEC basically takes total primary consumption of a nation and divides it by the population of that nation. It is not necessarily indicative of the actual energy use of individual end-users, but rather divides all of the energy use of all sectors and distributes it over the population number. Its usefulness is in comparing energy consumption patterns against time and against other countries.

Table 2.8 shows that the top five primary energy consumers in 2016 were China, the United States, India, Russia, and Japan. Dividing total energy consumption by population, we also see that PCEC in these five countries reveals very large discrepancies, particularly for the densely populated countries of China and India, which have very low PCEC compared to their high energy consuming counterparts. Should we expect these two countries to increase their per capita consumption moving forward? And if so, with such significant populations, how will that impact total global energy? And what role will energy efficiency play in dampening increasing PCEC?

China, for example, is a fascinating case when looking just 16 years in the past. In 2000, PCEC was a meager 0.8 TOE compared to the 2.21 TOE in 2016. PCEC increased by 2.76 times while total consumption increased by 3.03 times in that time frame. It is likely that their consumption will continue to grow for

Table 2.8 Energy Consumption and Population Data for the Five Largest Global Energy Consumers in 2016

Country	Total Energy Consumption (MTOE)	Population	Per Capita Energy Consumption (TOE)
China	3052.98	1,381,438,914	2.21
United States	2272.68	323,283,073	7.03
India	723.90	1,316,181,818	0.55
Russia	673.94	144,312,634	4.67
Japan	445.26	126,854,701	3.51

Figure created using data presented in British Petroleum's Statistical Review of World Energy (2017).

the foreseeable future. (For comparison, the US PCEC declined from 8.19 TOE in 2000 to 7.03 TOE in 2016. Both PCEC and total consumption declined in that time.)

2.5 DEFINITIONS AND REPRESENTATIONS OF ENERGY EFFICIENCY

We conclude this chapter by talking briefly about the general representation of energy efficiency and how it is applied to common processes and applications. As we have already discussed, efficiency is a ratio of energy input to the useful work achieved from that energy use. Since it is a ratio of two factors that have the same units, it has no units. And since the numerator is useful work (or heat equivalent), conservation measures that alter the useful work produced do not improve the efficiency. The formula itself is as simple as it can be: merely a ratio. The complexities come in accurately characterizing the energy input and the useful work within each step of a process. This requires accurate definitions of the energy inputs and the work. This, in turn, requires clear definitions of system boundaries.

In general, when dealing with energy processes and the resulting data that is collected, energy professionals tend to broadly define any energy efficiency ratio with the symbol, η, where it is equal to:

$$\eta = \frac{\text{Useful output of a process}}{\text{Energy input into a process}}$$

The useful output is what we desire, and the energy input is the cost needed to produce that useful output. It is heating food or a living space. It is processing materials to produce finished product. It is transporting people and goods. It is lighting spaces and powering electronics.

The useful work in heating food is quite clear. Space heating can be more complex. How much of the space does one really wish to heat? Small space heaters are commonly spoken of as being inefficient, but if the useful work is delivering heat to an individual at a desk, the useful work is not necessarily heating the entire house. Similarly, task lighting is often a very targeted need, rather than lighting an entire room. The useful work in transportation is moving the people or the goods, not the vehicle itself. According to Patterson, there are obvious methodological considerations that go into how to use the efficiency ratio, related to thermodynamic characteristics, physical output, economic goals, and service needs. Indeed, in some instances, it may be more appropriate to provide a ratio that focuses on the efficiency of delivery.

For example, most of the efficiencies that concern the need for heat usually has you start off with converting chemical energy into thermal energy. This can be done directly by combustion, or indirectly by electrical resistive heating. With respect to combustion, you basically burn a fuel, derive heat from it and direct that heat to your application. Combustion efficiency is basically the ratio between the heat released and the heating value of the fuel:

$$\text{Combustion efficiency: } \eta = \frac{\text{Heat released}}{\text{Heating value of fuel}}$$

Whether you want to heat food, a room, or a water tank, if you are using a combustible fuel, the combustion process is basically the same; however, heating values and other properties (like moisture content) affect overall combustion efficiency. This is more of a technical concern. However, other differences arise from the delivery of this released heat, or if another energy conversion is necessary for a desired outcome that requires combustion as an input.

In the instance of cooking, you want to burn a fuel to raise the temperature of the raw food to an adequate level so that it is safe and enjoyable to eat. In that case, cooking efficiency, or rather, the delivery of a cooking service could be defined as

$$\text{Cooking efficiency: } \eta = \frac{\text{Cooking heat delivered at a specified temperature}}{\text{Energy input}}$$

If the energy input is the burning of charcoal on a grill or burning of gas on a cook stove, the energy input would be the heating value of each fuel, and the cooking efficiency would mostly be affected by the characteristics of that fuel, plus any design characteristics of the grill or stove that enhance heat delivery at the specified temperature. If the energy input is resistive heating from an electric range, any efficiency losses have already happened at the electric power plant, and heating value characteristics are not necessary to consider.

Water heating and space heating would operate under similar conditions as cooking; however, you must also consider where and how that hot water or heated air is being delivered, and how well designed the space is to keeping heat within it. In the case of heating a home, you often assess efficiency via an "annual fuel utilization efficiency," which accounts for combustion efficiency, heat losses in delivery, and startup and shutdown losses over the course of the year.

There are many other examples and applications where efficiencies are determined based on individual components or entire processes that will be discussed in later chapters.

REFERENCES AND FURTHER READING

Books and Technical Articles/Reports
Blok, K., & Nieuwlaar, E. (2016). *Introduction to energy analysis*. Taylor & Francis.
Boyle, G. B. (Ed.), (2004). *Renewable energy*. Oxford University Press.
Culp, A. W., Jr (1991). *Principles of energy conversion*. McGraw-Hill.
Ebenhack, B. W., & Martínez, D. M. (2013). *The path to more sustainable energy systems: How do we get there from here?* Momentum Press.
Fowler, J. M. (1975). *Energy-environment source book*. National Science Teachers Association, Energy-Environment Materials Project.
Hills, R. L. (1993). *Power from steam: A history of the stationary steam engine*. Cambridge University Press.
Hinrichs, R., & Kleinbach, M. (2012). *Energy: Its use and the environment*. Nelson Education.
Krigger, J., & Dorsi, C. (2004). *Residential energy: Cost savings and comfort for existing buildings*. Helena, MT: Saturn Resource Management.
National Research Council (NRC). (2000). *Cooperation in the energy futures of China and the United States*. National Academies Press.
Patterson, M. G. (1996). What is energy efficiency? Concepts, indicators and methodological issues. *Energy Policy, 24*(5), 377−390.
Romer, R. H. (1976). *Energy: An introduction to physics*. Freeman.
Smil, V. (2008). *Energy: A beginner's guide*. Oneworld Publications.
Energy Information Sources and Reports
British Petroleum (BP). (2018). *BP energy outlook 2017 edition*. Available from <https://www.bp.com/en/global/corporate/energy-economics/energy-outlook.html>.
British Petroleum (BP). (2018). *BP statistical review of world energy*. Available from <https://www.bp.com/en/global/corporate/energy-economics/statistical-review-of-world-energy.html>.
Energy Information Administration (EIA). (2018). *Electric power monthly. Capacity factors for utility scale generators not primarily using fossil fuels*. Available from <https://www.eia.gov/electricity/monthly/epm_table_grapher.php?t = epmt_6_07_b>.
Energy Information Administration (EIA). (2018). *Monthly energy review*. Available from <https://www.eia.gov/totalenergy/data/monthly/>.
Energy Star. (2018). *Room air conditioner buying guide*. Available from <https://www.energystar.gov/products/heating_cooling/air_conditioning_room>.
National Aeronautics and Space Administration (NASA). (2018). *Prediction of worldwide energy resources*. Available from <https://power.larc.nasa.gov/data-access-viewer/>.

Organisation for Economic Co-operation and Development (OECD). (2018). *OECD data: Primary energy supply*. Available from <https://data.oecd.org/energy/primary-energy-supply.htm>.

Other Online Resources

Allen, J. (2018). *Principles of energy conversion eBook*. Department of Mechanical Engineering, Engineering Mechanics. Michigan Technological University. Available from <http://pages.mtu.edu/~jstallen/courses/MEEM4200/MEEM4200.html>.

Autodesk. (2018). *Sustainability workshop: Building energy loads*. Available from <https://sustainabilityworkshop.autodesk.com/buildings/building-energy-loads>.

American Physical Society (APS). (2018). *Panel of public affairs. Energy units*. Available from <http://www.aps.org/policy/reports/popa-reports/energy/units.cfm>.

Cobb, J. (2005). *An energy primer for the AP Environmental Science student*. Available from <https://apcentral.collegeboard.org/courses/ap-environmental-science/classroom-resources/energy-primer-ap-environmental-science-student>.

Energy Education. ca Encyclopedia. (2018). *Total primary energy supply*. Calgary, AB. Available from <http://energyeducation.ca/encyclopedia/Total_primary_energy_supply>.

Engineering Toolbox. (2018). *Higher calorific values*. Available from <https://www.engineeringtoolbox.com/fuels-higher-calorific-values-d_169.html>.

European Nuclear Society Encyclopedia. (2018). *Coal equivalent*. Available from <https://www.euronuclear.org/info/encyclopedia/coalequivalent.htm>.

Pacific Northwest National Laboratory (PNNL). (2018). *Hydrogen tools. HyARC calculator tools*. Available from <https://www.h2tools.org/hyarc/calculator-tools>.

Sherman, D. (2016). Horsepower vs. torque: What's the difference? *Car and Driver Magazine Blog*. Available from <https://www.caranddriver.com/news/horsepower-vs-torque-whats-the-difference>.

Society of Petroleum Engineers (SPE). (2007). *Petroleum resources management system*. Available from <http://www.spe.org/industry/docs/Petroleum_Resources_Management_System_2007.pdf>.

World Energy Council (WEC). (2013). *World energy resources 2013 survey (solar)*. Available from <https://www.worldenergy.org/wp-content/uploads/2013/10/WER_2013_8_Solar_revised.pdf>.

Primary energy trends

This chapter discusses global distributions and trends in energy resources. Because efficiency is a means to reduce consumption, the factors that guide supply and production patterns are critical. Key points include the following:

- Conceptualizing primary energy supply from a stock and flow perspective.
- Conceptualizing energy transformations from source to end use.
- Presenting historical and current trends in global energy resource supply and consumption.

3.1 CONCEPTUALIZING PRIMARY ENERGY SUPPLY

The global energy supply is commonly divided into two categories: "renewable" and "nonrenewable." Nonrenewable energy sources are classified as the fossil and nuclear fuels, coal, oil, gas, and uranium, while the renewable fuels are classified as hydropower and "new" renewable energy including wind, solar, geothermal, and biofuels.

As we have discussed in other works, it is unfortunate that primary energy has been classified this way, because it incorrectly implies that the resources in these categories are either limited (nonrenewable) or limitless (renewable) in their

Energy Efficiency. DOI: https://doi.org/10.1016/B978-0-12-812111-5.00003-2

utilization within the global supply. Supply is ultimately a physical limitation of how much energy is produced and efficiently used by human-controlled systems. And since supply is physical, each energy source has ultimate physical limits, either due to the finiteness of the resource or the finiteness and inefficiency of the infrastructures and technologies used to harness the energy from the resource.

Moreover, others have argued, and we agree, that energy sources are better described as being limited either by their stocks or by their flows. For example, fossil fuels and uranium are better described as the world's stock-limited energy supply. Our consumption of these resources is controlled by the total amount of mineral deposits known to be "in the ground" and available for exploitation, as well as the technical, economic, and environmental factors that allow for their practical and continued extraction.

As we explained in Chapter 2, Dealing with Energy Units, Measures, and Statistics, coal, oil, gas, and uranium all have tabulated reserves, meaning that the known available, exploitable mineral stock of each resource is immediately available to use. These stocks are rather sizeable, and because they are well established in the global resource supply infrastructure and are the basis for how we have created machines to serve our end uses, they have allowed for a high degree of predictability for the global economy. We can keep the "lights on" and move about all the time because these resources are exceedingly abundant, highly transportable, and are available to be exploited at any time.

As depicted in Fig. 3.1, we can also add to the reserves supply from the estimated resource base by making new discoveries, or by improving recovery via better engineering technologies and practices. We can also extend reserves availability by consuming less via efficiency enhancements or conservation practices. In the end, however, since the total resource stock is not renewed (at least not on time frames equal to how we have been consuming them), it will not be available indefinitely. Moreover, the faster we extract them for our needs and wants, the sooner that supply will be depleted; thus, they are stock-limited.

In contrast, solar and wind energy are best described as a flow-limited energy supply. Our consumption of these resources is controlled by the flow of sunlight

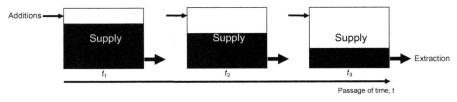

FIGURE 3.1

Depiction of stock-limited energy over time. The supply continues to be reduced as it is being extracted for use. The supply will dwindle as quickly as it can be extracted from the total resource base. Reserves additions and reduction in extraction rates will extend the lifetime of the exploitable stock, but the stock will ultimately deplete.

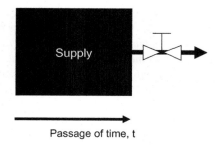

FIGURE 3.2

Depiction of flow-limited energy. The energy supply will never deplete over time; however, the flow of energy is controlled by an incident flow rate, represented by the valve.

or flow of wind energy onto the energy collector (e.g., a panel or a turbine). There are no reserves of sunlight or wind from which to draw, but rather the energy must be used and/or converted as it is collected from its practically nondepletable supply (i.e., the sun or wind); however, it is limited to the incident flow rate, as depicted in Fig. 3.2.

The valve used in Fig. 3.2 is meant to convey that when the sun is not shining or the wind is not blowing sufficiently over Earth-based collectors, no substantial energy will be available to utilize. Thus, it is limited by flow. The benefit of relying on solar and wind energy flows is that the amount of energy used at any given time has no direct effect on how much is available in the future. That is, the supply does not change. The penalty of using this type of supply, however, is that we are subject to the relative availability of the originating supply at any given moment. If the sun is not shining or the wind is not blowing, we cannot use them directly.

Other resources, such as traditional, dam-based hydropower, biomass, and geothermal energy are best classified as stock- and flow-limited energy. Traditional hydropower, for example, taps into the flow of water, constantly renewed by the hydrologic cycle, but dams can be built to stockpile some of the sources, for use on demand. Similarly, biomass is derived from photosynthesis, which is also a continuous process, and forests are the stocks of existing biomass. However, it is possible to overconsume and deplete the energy stored in the biomass stock by cutting down too many trees too quickly before regrowth can occur. It is also possible to erode soils, which are needed for sustained regrowth of biomass. Likewise, geothermal reservoirs can be depleted more quickly than they can regenerate, which requires a delay in extraction.

Again, the type of source affects both how the energy is acquired and what form it takes. All of the stock-limited energy sources, such as the fossil fuels, are extracted from existing, natural storage at rates based on physical factors and local, regional, and global demand. The flow-limited resources, such as solar and wind, can only be harvested locally at the rates of their natural flow. That is, the energy is converted to its final form at the point of capture. Since the source is not always available on demand, these flow-limited resource systems often require battery storage to meet demand when needed, if they are providing a large share of the total power—which is the case for a stand-alone solar or wind power system. The many sources that have aspects of both flow and stock limitations tend to use their stock characteristics so that energy can be produced at rates determined by demand.

Finally, most "new" renewable technologies are being employed to generate electricity. The current model in developing these energy systems is based on the large-scale centralized power generation model that emulates coal and nuclear power systems. This incurs the efficiency losses associated with long distance electric power transmission. This is particularly true for wind because the windiest sites are not typically located near large population centers. Although sunshine is more appealing, the sunniest sites must have relatively little cloud cover—and little rain. The Sahara, for example, offers great solar potential but is remote from population centers.

As we consider the availability of energy resources and their use in the 21st century, as well as the role of energy efficiency in utilizing and extending resource lifetimes, we must keep in mind these two primary categories of stock and flow limitations.

3.2 CONCEPTUALIZING PRIMARY ENERGY FLOWS TO END USE

Once an energy source is determined to be available for productive end use, in the form of fuels, heat, and/or electricity, it is then converted from the available supply to ongoing production. The stock-limited fossil and nuclear fuel sources continue to produce from a respective reservoir until some physical, economic, and/or technical constraint is reached. Flow-limited sources, on the other hand, continue to produce indefinitely, albeit intermittently. Both types of supply depend on adequate operation and maintenance to ensure production throughout the lifetime of the equipment needed to extract the primary energy supply and convert to useful, productive energy.

All fossil fuels are found in the Earth's subsurface and are acquired through mining or drilling. Coal begins with mining, while oil and gas begin with drilling, if one discounts the initial exploration steps.[1] These processes are quite energy

---

[1]There is a common misperception that modern technology enables exploration to be done from the surface, before drilling wells. However, there are many variables in the subsurface, and the resource is often separated from the surface by more than two miles of rock. We do not know what is far below the Earth's surface before drilling into it, or sinking a mine shaft.

intensive, which severely impact the overall system efficiency but are nonetheless small compared to the immensity of the energy recovered from successful fossil fuel producing fields. For oil and gas drilling or for coal mining, the energy input is exponentially correlated with the depth of the resource. Not only is more energy naturally required to drill or dig deeper but also the equipment required to accomplish it is larger and more powerful and the rocks penetrated are likely to be harder. The likelihood of failure means that some wells and mine shafts will not produce. These "dry holes" and other failures count against the energy efficiency of overall acquisition operations but are still overwhelmed by the enormity of successful production.

Once drilled or mined, these three resources enter the supply, production, and distribution stream following a relatively straightforward path to end-use energy, as depicted in Figs. 3.3–3.5.

Looking at energy flow from source to end use in Figs. 3.3–3.5, we notice that after mining/drilling there can be some fuel processing and then transport of fuel to the consumer, which can be a power plant for electricity production, an industry for energy-intensive manufacturing, or residential and commercial end users of fuels, heat and/or power plant electricity. The processes in Figs. 3.3–3.5 represent examples based on the dominant uses of each source. Coal, depicted in

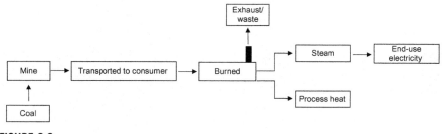

FIGURE 3.3

A simplified depiction of the flow of energy from source to end use of the coal fuel system.

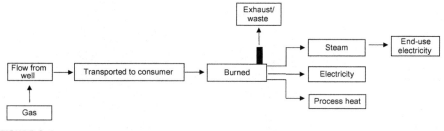

FIGURE 3.4

A simplified depiction of the flow of energy from source to end use of the natural gas fuel system.

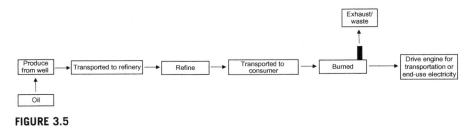

FIGURE 3.5

A simplified depiction of the flow of energy from source to end use of the oil fuel system.

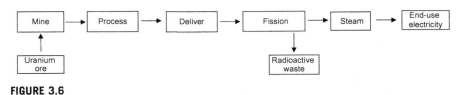

FIGURE 3.6

A simplified depiction of the flow of energy from source to end use of the nuclear fuel system.

Fig. 3.3, is used extensively for both process heat and for electric power generation. Also, note the absence of a processing step: coal may be crushed and dried but can also be burned directly. Gas, depicted in Fig. 3.4, also lacks a processing step, since it tends to be a very clean, efficient fuel as it comes out of the reservoir. It is often run through a simple separator to remove water vapor and may be treated to remove hydrogen sulfide contamination if it is "sour." Otherwise, it can be delivered directly to the consumer. Oil, depicted in Fig. 3.5, is refined, not because it cannot burn directly, but because it is processed to make precisely engineered fuels to be used for specific engine designs. This adds an extra shipping step to and from the refinery, before distribution to consumers. The liquid fuels (e.g., jet fuel, diesel, gasoline, etc.) that can be refined from crude oil use the majority of the oil, largely because their energy densities make them highly transportable fuels for mobile applications.

Similarly, uranium ore must be mined and processed to deliver usable fuel. The supply, distribution, and end use follow paths akin to the fossil fuels, as depicted in Fig. 3.6.

Nuclear fuel, used for nuclear electric power, begins with the fission of unstable nuclei of the relatively rare isotope of uranium, U-235. All uranium isotopes naturally and spontaneously undergo decay, releasing energy as radiation and, ultimately, heat. Nuclear power is always converted to heat, but for global energy use the thermal energy derived is essentially always converted to kinetic energy to drive a turbine to generate electricity, never to drive a piston, nor directly for process heat. The energy conversion is too intense to be used effectively for the lower level activities.

The overall process of ore to electricity is rather involved, because uranium processing is a major, essential step, and far more energy intensive than fossil fuel processing. Even very pure uranium must be enriched to be useful as a fuel. Since the enrichment step means separating two isotopes of the same compound, which are nearly identical in every way, it requires complex processing: reacting the uranium with hydrofluoric acid to make uranium hexafluoride gas, which can be partially separated in a centrifuge, based on the small difference in the mass of the uranium isotopes.

Due to the unparalleled energy density of nuclear reactions, very little fissile material is required in a nuclear power plant: about 6.5−9 MT/year of operation of a 1-GW reactor, depending on the enrichment level. This means that much less ore is required than coal to produce the same amount of electricity. The largest penalty of using nuclear fuels (in addition to the highly energy intensive and inefficient processing step) is the creation of long-lived radioactive waste, which must be managed for decades or longer. However, it is possible to tap a significant portion of the radioactive waste with breeder reactor technologies.

In contrast, most renewable energy resources originate above Earth's surface and their flows are acquired or harvested by resource-specific collectors. As mentioned above, because societies need reliable energy to function, these flow-limited resources are often coupled to storage or tied directly to existing energy infrastructures (i.e., the electric grid) and other technologies.

Solar energy, for example, can be collected actively for electricity production by concentrating the energy onto collectors that then generate pressurized steam for use with turbine/generator sets (i.e., solar thermal electric power), or they can be collected using photovoltaic (PV) technology that converts photons into electricity.[2] These two types currently dominate the recent global energy supply additions from the solar resource and the solar energy flows from source to electricity end use is depicted in Figs. 3.7 and 3.8.

Since electricity is energy in flux, the only way to have the energy available on demand is to store any excess converted electricity for future use through some form of storage. The prevailing technology is via electrochemical storage (a

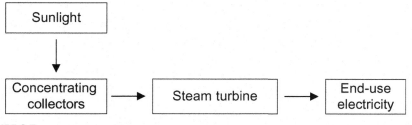

FIGURE 3.7

Simplified depiction of the flow of energy from source to end use of the solar energy supply to electricity for solar thermal electric power technologies.

---

[2]Solar energy can also be collected passively for natural lighting and for building heating needs.

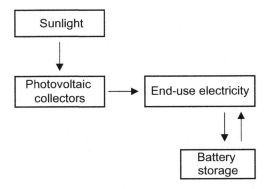

FIGURE 3.8

Simplified depiction of the flow of energy from source to end use of the solar energy supply to electricity for solar photovoltaic technologies.

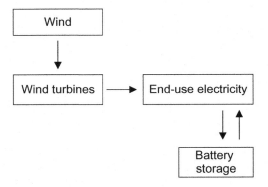

FIGURE 3.9

Simplified depiction of the flow of energy from source to end use of the wind energy supply to electricity for wind turbine technologies.

battery), which is limited by the capacity of each battery cell. Energy densities of common batteries range from below 50 Wh/kg for lead—acid battery systems to about 200 Wh/kg for modern lithium polymer battery systems. That is equivalent to 0.18—0.72 MJ/kg. When compared to the energy content of the fossil fuels, which range from 20 to 40 MJ/kg, it is clear that battery technology will need to advance further for the solar resource to be used in an equivalent manner.

Similar to solar PV, wind energy has direct conversion of the primary source to electricity, as depicted in Fig. 3.9. Place a wind turbine in a sufficiently windy location and electricity can be produced. Turbines can be employed in "off-the-grid" applications, for which battery and/or other types of storage would be necessary, but a major share of investment in wind power systems is for large, utility-scale generation that feeds directly into the grid.

As we explained in Section 3.1, hydropower is best classified as both stock- and flow-limited energy. Water can be stored behind dams, and the stored potential energy can be released on demand to flow through water turbines to produce electricity. This process is extremely effective, thus most of installed hydropower capacity employs large dams to concentrate and store the energy of the water, as depicted in Fig. 3.10.

Due to this ability to hold water in reservoirs for controlled energy production, as well as other ancillary benefits of controlling water supply, hydro dams have been extensively utilized across the world. Because of this and because conversion efficiency is very high, hydropower has thus far dominated the total renewable energy supply.

Biomass, another stock- and flow-limited energy source, continues to be utilized extensively throughout the world. Forests and agricultural biomass store chemical potential energy much like the fossil fuels but can regrow at rapid enough rates to replenish stocks in a more reasonable time frame (months to decades). Much of the world continues to use firewood and processed charcoal for heating and cooking and the flow of energy from source to end use is depicted in Fig. 3.11.

Modern biofuels, such as biodiesel derived from oil producing seeds, or fuel alcohol from sugarcane or corn crops, have been growing in utilization in many developed and developing countries. Also, many nations have begun mixing

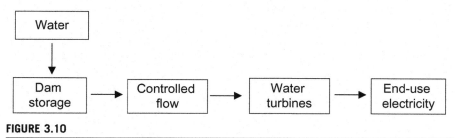

FIGURE 3.10

Simplified depiction of the flow of energy from source to end use of the hydropower supply to electricity for water turbine technologies.

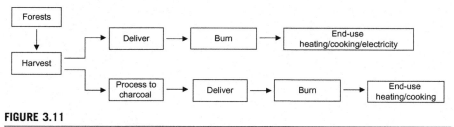

FIGURE 3.11

Simplified depiction of the flow of energy from source to end use of the forest biomass supply to end use.

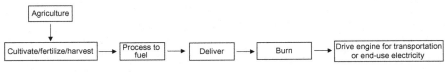

FIGURE 3.12

Simplified depiction of the flow of energy from source to end use of the agricultural biomass supply to end use.

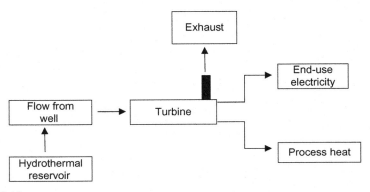

FIGURE 3.13

Simplified depiction of the flow of energy from source to end use of the geothermal supply to end use.

woodchips or sawdust derived from fast-growing trees planted and harvested very similarly to agricultural crops burned for electricity production. The flow of energy from these sources to end use is depicted in Fig. 3.12.

Geothermal energy is yet another stock- and flow-limited energy source used widely around the world. A geothermal energy resource is simply a geologic, hydrothermal reservoir from which heat from the subsurface water or steam can economically be extracted, and is acquired through drilling, similar to oil and gas. Geothermal energy can be used for both electricity conversion and process heat, including direct use applications such as district heating for municipal and residential needs. The flow of energy from source to end use is depicted in Fig. 3.13.

Conceptually understanding the flow of source to end use is important for analyzing energy flows through economic sectors, which is the focus of Chapter 4, Energy Flow Analyses and Efficiency Indicators.

3.3 OIL AND GAS SUPPLY AND PRODUCTION

Oil and gas are a collection of liquid and gaseous hydrocarbons held in vast geologic formations of porous, permeable rock, known as petroleum reservoirs. Oil and gas companies gain access to the hydrocarbons found in these reservoirs by drilling wellbores and investigating the potential for commercial viability to sell to local or international markets. Once a sufficient amount of oil and gas is discovered in a field or in a region of fields, production facilities are built.

Oil production can range from just a few barrels of oil per day in small fields to several thousand barrels of oil in large fields. The amount of energy contained within each barrel is substantial and the amount of oil available across the world remains massive. Indeed, as discussed in Chapter 2, Dealing with Energy Units, Measures, and Statistics, the amount of oil or gas is often described by their reserves and the estimates for global reserves is updated annually. In 2016, the proved global reserves of oil alone were estimated by British Petroleum (BP) in its 2017 *Statistical Review of World Energy* to be over 1.7 trillion barrels (10^{12} bbl). In terms of energy content, this is roughly equal to 10,000 quads or 10,000 EJ of energy stored in the world's oil reserves. (Note that the world consumed nearly 600 EJ in 2016.)

As shown in Fig. 3.14, current proved recoverable oil reserves are highly concentrated in the Middle East (nearly 48% of total); however, substantial increases in proved recoverable reserves have been achieved in other regions in recent times. There has been discussion of limits to petroleum production ever since "large-scale" commercial production began in the United States, with the famous "Drake" well in Pennsylvania.[3] Particularly throughout the latter half of the 20th century, and beginning of the 21st, there have been multiple projections of "running out of oil" in the next 20 years. Yet, petroleum production has generally continued to increase (recall Fig. 2.2, which showed the world produced a record 92.2 million barrels per day in 2016).

For examples of recent growth, Fig. 3.15 shows massive increases in proved recoverable oil reserves in Canada and the United States since the mid-1990s. Canada's increase was the result of production from new discoveries, while the United States increase can be directly attributable to advances in technology including hydraulic fracturing and horizontal drilling techniques. However, despite these massive increases in certain regions, some countries, such as countries in the EU, have very small reserves and thus are highly dependent on other countries for oil. It is unlikely that EU reserves will grow considerably, which at least partially indicates Europe's desire (and need) to move away from reliance on a small resource that they have little control over.

Since oil is traded in barrels, it is often reported how each barrel is refined to generate the fuels used by local markets. As displayed in Fig. 3.16, for example, gasoline and (low sulfur) diesel fuel represent the bulk of the refined products

---

[3]"Large scale" is, of course, a relative term. Drake's well barely produced 10 bbl/day, but that was huge in comparison to other oil sources of that time. Within a few years, the first single wells were drilled that could initially produce 1000 bbl/day. Petroleum production quickly dwarfed all other oil sources and continued to grow exponentially.

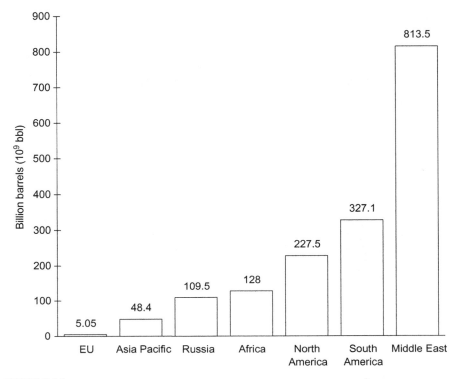

FIGURE 3.14

Proved recoverable oil reserves by region or country in units of billions of barrels, end of 2016.

Figure created using data presented in British Petroleum's Statistical Review of World Energy (2017).

produced from each barrel in the United States, roughly 69%. Gasoline is the preferred fuel in the United States and Asia. Europe has favored diesel fuel much more than gasoline in part because diesel engines are more efficient than gasoline engines. However, concerns over excess particulate pollution from diesel passenger vehicles in recent years are beginning to affect that trend.

Natural gas has, likewise, undergone a similar expansion in reserves across the world in recent years. Fig. 3.17 shows that natural gas reserves increased from 119.9 trillion cubic meters in 1995 to 186.6 trillion in 2016 (recall $1 \text{ m}^3 - 35 \text{ ft}^3$). In terms of energy, roughly 7000 EJ of energy is stored in the world's gas reserves.

Annual oil and gas production growth remains strong due to an extreme dependence on oil for transportation and a rapidly expanding demand for gas to generate cleaner burning electricity. (Natural gas has also become cheaper, thanks to the dramatically increased supply of gas produced from shales in recent years.) Indeed, oil and gas together accounted for more than 57% of the world's 2016 primary energy supply that was used to drive the global economy.

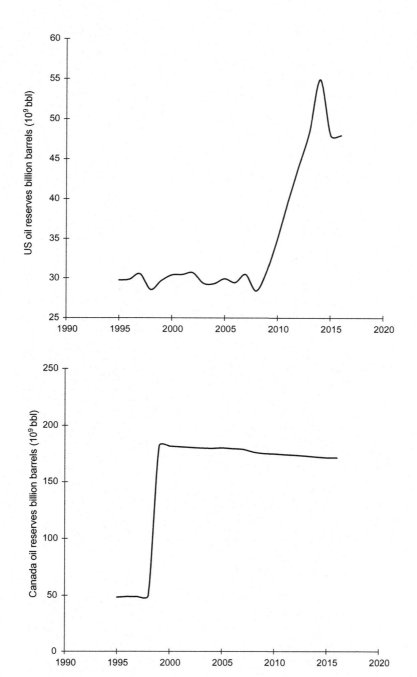

FIGURE 3.15

Proved recoverable oil reserves for Canada and the United States since 1995 in units of billions of barrels.

Figure created using data presented in British Petroleum's Statistical Review of World Energy (2017).

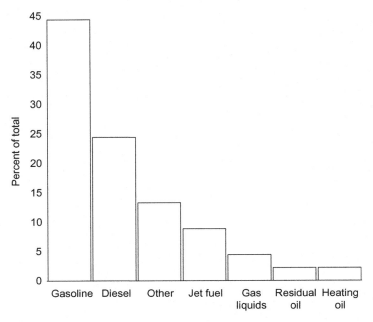

FIGURE 3.16

Percentage distribution of refined fuels made from 1 bbl of crude oil in the United States. Forty-four percent of a barrel is used for producing gasoline.

Figure created using data presented in the US Energy Information Administration (EIA) Petroleum Supply Monthly (2018).

3.4 COAL SUPPLY AND PRODUCTION

Coal forms under several conditions, but swamps represent one of the most common sources. The growth, death, and burial of raw biomass into stagnant waters permit accumulation and decay without oxidation. The carbon, then, remains fixed in place. These conditions can result in the creation of thick layers of peat, and if eventually buried under rock sediments, it will likely result in the creation of coal (albeit over thousands to millions of years). Unlike in oil and gas, coal forms in solid beds, not in voids of porous rock.

Young coals are soft and are often called "low rank coals," such as lignite, and look much like peat or the trees from which they are largely derived. The oldest coals are often called "hard coals" and are the bituminous and anthracite varieties. The more mature coals are, the cleaner and hotter they burn, producing very little smoke.[4] However, the most abundantly available ranks of coal tend to be

---

[4]The decaying organic matter is buried under stagnant water and initially contains a great deal of moisture, but water is squeezed out over geologic time, causing more mature coals to have lower moisture content, which is a large part of why they burn hotter and, thus, cleaner.

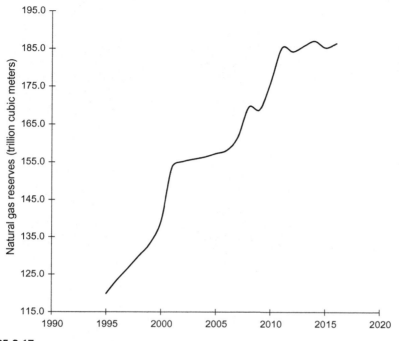

FIGURE 3.17

Proved recoverable natural gas reserves for the world in units of trillion cubic meters 1995–2017.

Figure created using data presented in British Petroleum's Statistical Review of World Energy (2017).

subbituminous and bituminous, which are used for electricity generation, as well as for industrial materials, such as for making iron, steel, and cement. The percentage abundances and most common uses as reported by the World Coal Association are presented in Fig. 3.18.

The largest coal mines in the world can produce between 10 and 100 million tonnes annually, and the distribution of coal reserves of all types by region is presented in Fig. 3.19.

Total world coal reserves at the end of 2016 stood at 1,139,331 million tonnes. This is equivalent to 45,000 quads or 47,000 EJ of energy stored in the world's coal reserves.

Globally, coal is used primarily for electricity generation and for aluminum and steel production, and secondarily for residential and commercial heating especially in China. However, with a push to reduce coal consumption for electricity in favor of natural gas, solar, and wind, as well as a concerted desire to reduce the energy intensity of metals manufacturing, there has been an extremely large reduction in coal production in the past 5–7 years. Again, the bulk of this reduction has been because of Chinese government mandates to reduce coal use in the electricity and residential sectors in favor of natural gas, as well as a reduction in coking coal for metals. As can be seen in Fig. 3.20, world coal consumption has fallen dramatically in recent years.

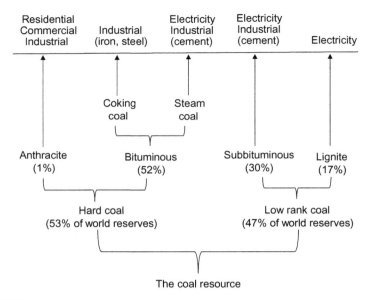

FIGURE 3.18

Coal types, abundances, and uses.

Figure created using data and imagery presented in the World Coal Association's Types of Coal (2018), with adaptation.

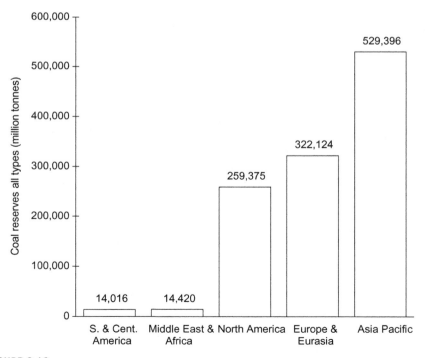

FIGURE 3.19

Coal reserves of all types by region in units of millions of tonnes.

Figure created using data presented in British Petroleum's Statistical Review of World Energy (2017).

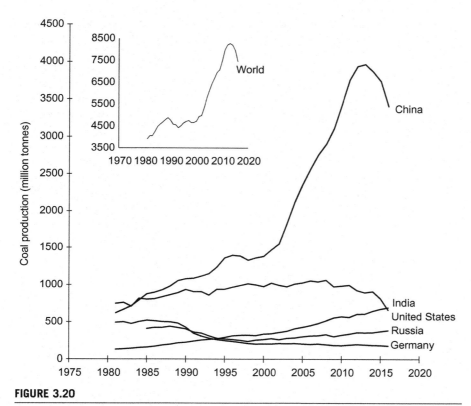

FIGURE 3.20

Coal production from 1980 until 2016 by the world and the top five producers. Notice significant reductions in production by the United States and China in the past 5–7 years. Also notice that India and the United States are producing almost the same amount; however, India has been increasing production, while the United States has been decreasing production.

Figure created using data presented in British Petroleum's Statistical Review of World Energy (2017).

It is likely that coal will continue to be utilized quite significantly both for electricity and manufacturing applications; however, in spite of the massive reserves, paradoxically, the massive reductions in production means its dominance in these sectors appears to be diminishing. Between energy efficiency improvements and fuel switching, it is quite probable that this downward trend will continue. However, the fact that 47,000 EJ of energy remain in the ground and the fact that there exists roughly 2000 GW of installed, operating electricity generating capacity from coal, suggests that coal will continue to be utilized well into the latter half of this century.

3.5 URANIUM SUPPLY AND PRODUCTION

Uranium is relatively abundant in Earth's crust and can be found in most rock types, sands, and even in seawater. A concentrated uranium orebody is defined by the World Nuclear Association as "an occurrence of mineralization from which the metal is economically recoverable." Uranium, probably more than oil, gas, or coal, needs favorable economics to be considered recoverable. That is, it needs to have low costs of extraction and high market prices. Thus, the known recoverable uranium resources found around the world are sensitive to specific monetary thresholds. With those caveats, the most recently reported (2015) known recoverable resources of uranium were 5,718,400 tonnes, and the percentage distribution by country is found in Table 3.1. This number was based on a market value of US $130/kg of uranium.

As mentioned in Section 3.2, nuclear fuel is used for nuclear electric power exclusively, that is, nuclear fuel is used to generate steam for turbine/generator sets across the world. More than 30 countries use nuclear energy for electricity production in about 450 nuclear reactors. In 2016, the installed nuclear power plant capacity was about 392 GW, and an additional 60 plants with an additional 60 GW were under construction.

Table 3.1 Known Recoverable Uranium Resources 2015

Country	Tonnes of Uranium	Percentage of World Total
Australia	1,664,100	29
Kazakhstan	745,300	13
Canada	509,000	9
Russian Fed	507,800	9
South Africa	322,400	6
Niger	291,500	5
Brazil	276,800	5
China	272,500	5
Namibia	267,000	5
Mongolia	141,500	2
Uzbekistan	130,100	2
Ukraine	115,800	2
Botswana	73,500	1
The United States	62,900	1
Tanzania	58,100	1
Jordan	47,700	1
Other	232,400	4
World	5,718,400	100

Table created using data presented in the Organisation for Economic Co-operation and Development's Nuclear Energy Agency Uranium 2016: Resources, production, and demand (2016).

As seen in Fig. 3.21, BP reports that global electricity consumption from nuclear fuel increased steadily from 1965 to about 2005 and plateaued to about 2500 TWh until the 2011 Fukushima Daiichi nuclear disaster in Japan, which resulted in a 100% closure of nuclear power in Japan. This, coupled with several decommissioned reactors in Germany, resulted in a severe drop in nuclear powered electricity production (and, thus, consumption) in a fairly short time frame. However, Japan has begun to reopen reactors, which, along with massive expansion of nuclear power in China, has led to a rebound in nuclear generated electricity in recent years. Note that because nuclear electric power plants are thermal electric plants, they require more thermal input to produce the 2500 TWh of electrical energy. Assuming a 38% thermal plant efficiency, 6579 TWh of thermal energy from nuclear fission is required to produce the 2500 TWh of electrical energy (2500 TWh is equivalent to 9 EJ of electrical energy produced from nuclear power).

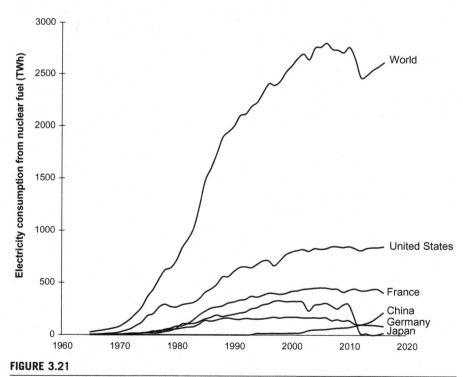

FIGURE 3.21

Electricity consumption from nuclear fuel from 1965 until 2016 from the world and other notable consumers. Notice significant reductions in consumption by Japan and Germany since 2011 juxtaposed with significant increase in consumption by China.

Figure created using data presented in British Petroleum's Statistical Review of World Energy (2017).

3.6 HYDROPOWER SUPPLY AND PRODUCTION

Hydropower remains the most widely utilized renewable energy source globally. The International Hydropower Association (IHA) reports a total installed world hydropower capacity of 1245 GW in 2016. The top five countries with the most installed capacity are depicted in Fig. 3.22. China has the most installed capacity primarily because of recent capacity additions. In terms of actual use, global electricity consumption from hydropower was just over 4100 TWh in 2016, with historical consumption displayed in Fig. 3.23 from 1965 to 2016. The steeper slope from about 2000 is, again, almost completely due to large increases in hydropower additions in China.

Despite most of the best sites already being developed for large-scale hydropower production, it is estimated that the world annual, technically exploitable capability from hydropower sources are roughly 17,500 TWh (about 63 EJ).

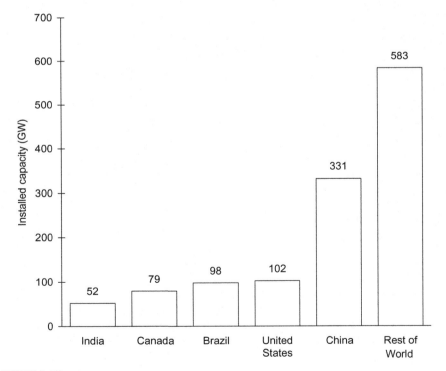

FIGURE 3.22

Installed electricity generating capacity from hydropower in units of GW in 2016 for the top five producers and the rest of the world.

Figure created using data presented in the International Hydropower Association's Hydropower Status Report (2017).

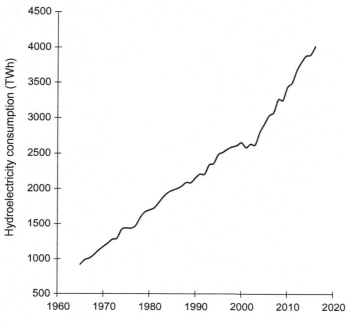

FIGURE 3.23

Global electricity consumption from hydropower from 1965 to 2016.

Figure created using data presented in British Petroleum's Statistical Review of World Energy (2017).

Environmental restrictions may ultimately inhibit expansion to this technically exploitable amount; however, the potential for expansion is likely, based on the desire for the consumption of low carbon energy[5].

It is clear that hydropower production capacity and consumption have been growing steadily over the half century represented in the data shown. There can be no doubt that a limit to the growth of traditional riparian large-scale hydropower will be seen.

3.7 WIND SUPPLY AND PRODUCTION

Harnessing wind energy involves the passing of flowing air through a wind turbine blade/generator set, much like harnessing water energy with a water turbine/generator set. The most important difference between the two fluid

---

[5]Although most of the large-scale hydropower is currently produced from "big dam" facilities on rivers, projects are underway on tapping tidal, current, and wave (other forms of hydropower) at a large scale. The success of these efforts could radically transform our understanding of remaining large-scale hydropower potential.

harvesting systems is density and fluid storage. Air is about 800 times less dense than water and cannot be stored behind a dam. Also, whereas the hydrological cycle is the primary source of water movement throughout the world, surface temperature and pressure differences are mostly responsible for wind movement.

The primary energy conversion for harnessing the wind is kinetic energy to electrical energy, and the two largest factors for maximizing electricity production from the wind resource are the swept area (i.e., the size of the wind-blown area) and, more importantly, the air velocity (i.e., wind speed). In fact, while wind power is proportional to the swept area, it is proportional to the cube of the wind speed.

Wind resources are not distributed equally across the world. Regional average speeds range from below 2.5 m/s in the least windy regions to greater than 9.75 m/s in the windiest, with equivalent power densities of less than 25 W/m^2 to greater than 1300 W/m^2. It is notable that wind speed is stronger high above the immediate surface, due to reduced frictional barriers from natural and human-made obstructions; thus, there has been a concerted effort to measure wind speeds at varying heights, as high as 200 m above the surface. Wind is obviously an intermittent energy source; thus, average speeds do not directly correlate with constant, available energy.

Advances in turbine blade and tower design since the mid-1990s have led to a massive increase in global installed wind power capacity. From 1997 to 2016, BP reports that installed capacity grew from 7.6 to 469 GW.[6] For reference, Fig. 3.24 shows installed wind capacity of the five largest wind electricity generating countries from 1997 to 2016. China has been by far the largest investor of wind energy since 2010.

In terms of actual use, global electricity consumption from wind power was just under 1000 TWh in 2016, with historical consumption, 1987−2016, displayed in Fig. 3.25. Based on an installed capacity of 469 GW in 2016, this indicates a wind capacity factor of about 0.24. (Capacity factor was defined in Chapter 2, Dealing with Energy Units, Measures, and Statistics.)

Finally, estimated annual technical potential for wind is on the order of hundreds of thousands of terawatt-hours per year (100,000 TWh ∼ 360 EJ). Though it is unlikely that all, or any large share, of this potential will be realized, continued improvements in technology make it likely that wind energy expansion will continue into the foreseeable future. It does offer low carbon energy, whose future availability cannot be diminished by its current use.

3.8 SOLAR SUPPLY AND PRODUCTION

As we explained in Chapter 2, Dealing with Energy Units, Measures, and Statistics, the sun is continuously bombarding a flat surface of Earth's atmosphere

---

[6]We are using 2016 as the end case when possible in this chapter; however, it is notable that the World Wind Energy Association reports an additional 52.6 GW of new installed capacity was added in 2017.

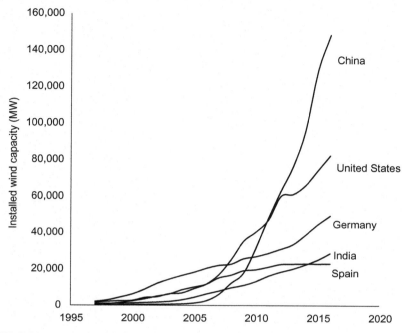

FIGURE 3.24

Installed wind capacity additions for the top five wind electricity generating countries from 1997 to 2016. China had nearly twice as much installed capacity as the next largest in 2016.

Figure created using data presented in British Petroleum's Statistical Review of World Energy (2017).

with electromagnetic radiation at a rate of roughly 1370 W/m^2. After accounting for Earth's geometry and atmospheric variables, this value decreases to an average value of about 170 W/m^2. This is the commonly accepted average annual value incident on Earth's land and ocean surfaces, defined as "insolation."

A common academic exercise is to estimate the amount of power that could be converted to electricity, assuming a certain conversion efficiency. For example, assume that we have the technological means to convert 25% of solar insolation into electricity and we had access to all of the land on the planet. Earth's land surface area is roughly equal to 1.5×10^{14} m^2, and the annual insolation average is 170 W/m^2; thus, solar insolation on land would roughly equal 2.5×10^{16} W. Assuming 25% solar to electric conversion efficiency (i.e., 1 W of solar produces 0.25 W of electricity), humans could theoretically generate 6.25×10^{15} W or 6.25 million GW of electricity, annually. That is about 1000 times more than the currently installed electricity capacity, so it should be clear that the resource is exceedingly abundant. Of course, at specific times of the day and year, and at specific latitudes, solar insolation at any given location can vary widely from that average value. Like wind, it is an intermittent resource.

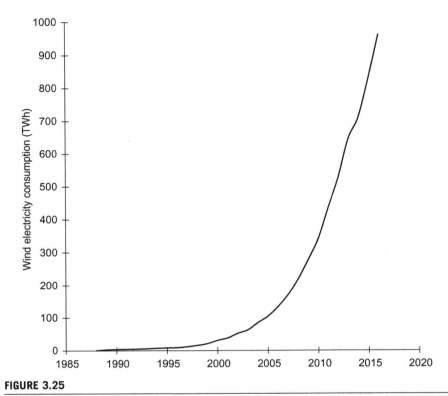

FIGURE 3.25

Global electricity consumption from wind power in units of TWh from 1987 to 2016.

Figure created using data presented in British Petroleum's Statistical Review of World Energy (2017).

In terms of major global production, solar electric power is primarily acquired through two paths: either concentrating solar power (CSP) to make steam to drive turbines or using solar PV cells. The PV cells directly convert light energy to electricity, while solar thermal applications require an additional step (or two, if you consider boiling water and running a turbine as separate steps).

CSP technologies, including parabolic troughs, linear Fresnel reflectors, power towers, and dish/engine systems, are dependent on locations with consistent, intense sunlight. Deserts are often prime locations. Because of this, CSP is not broadly utilized. In 2016, CSP had a total operational capacity of 5.2 GW led by the United States, Spain, South Africa, Morocco, and India. An additional 4.7 GW is under development or under construction in other regions.

On the other hand, capacity additions from solar have mostly come from massive investment in residential and utility-scale solar PV projects. As depicted in Fig. 3.26, since 1997, installed capacity has dramatically increased from just under 0.3 GW to over 300 GW in 2016. Reduced manufacturing costs and extremely favorable industry incentives are equally responsible for this astonishingly rapid increase (representing slightly more than 40%/year over a 20-year

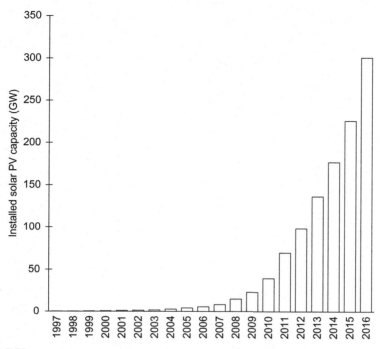

FIGURE 3.26

Global installed solar PV capacity in units of GW from 1987 to 2016.

Figure created using data presented in British Petroleum's Statistical Review of World Energy (2017).

time period). China, the United States, and Germany were the top three countries with new capacity additions in 2016.

In terms of actual use, global electricity consumption from solar power was about 333 TWh in 2016 (both CSP and PV), with historical production (and, thus, consumption) displayed in Fig. 3.27 from 1989 to 2016. Based on an installed capacity of 306.2 GW in 2016, this indicates a solar capacity factor of about 0.124. Compare this to 2500 TWh produced from nuclear with a similar installed capacity. This low capacity factor indicates the relatively small contribution solar currently makes to global electricity production. With increased environmental concerns from combustion-based electricity generation, decreased prices of solar installation, and increased incentives for residential generation, the contribution will likely increase.

3.9 BIOMASS SUPPLY AND PRODUCTION

Biomass is all growing plant matter, defined by the US National Renewable Energy Laboratory as "trees, grasses, agricultural crops, or other biological

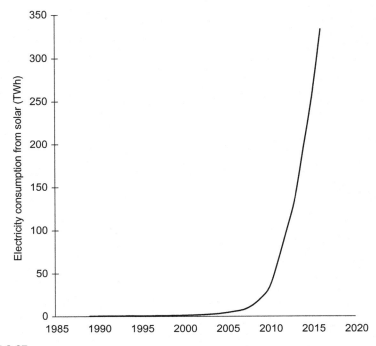

FIGURE 3.27

Global electricity consumption from solar power from 1989 to 2016.

Figure created using data presented in British Petroleum's Statistical Review of World Energy (2017).

material." All these materials can be burned for energy, and humans have long used biomass energy as a major primary energy source. Indeed, it is only recent in the history of humanity that other energy sources have taken over the primary share of total energy supply.

Raw, woody biomass is commonly harvested from natural forest growth. It is used as a solid fuel for cooking and heating in the form of firewood, charcoal, and waste. (Solid fuel can also be derived from agricultural waste in the form of dung.) It is also converted into liquid or gas to produce more refined fuels for cooking, electricity, and even transportation. Modern, engineered biomass fuels can be derived from a number of energy crops, usually by pressing seeds to make oil fuels (primarily from soy) or by converting them into alcohol fuels (from corn or sugar cane). Ethanol and biodiesel are the two most widely produced biomass fuels utilized by international markets. Finally, municipal and industrial waste can also be burned for energy needs.

With all the attention that ethanol and biodiesel have received in recent years, it can be difficult to remember that solid biomass still is, by far, the dominant contributor to the total primary biomass energy supply. In fact, the primary energy supply from forest biomass was estimated to be roughly 53 EJ in 2014,

Table 3.2 The Global Biomass Energy Supply in Units of EJ From 2000 to 2014

Year	Solid Biomass (EJ)	Liquid Biofuels (EJ)	Municipal Waste (EJ)	Biogas (EJ)	Industrial Waste (EJ)
2000	41.10	0.42	0.74	0.28	0.47
2005	44.70	0.85	0.94	0.50	0.40
2010	49.10	2.44	1.15	0.84	0.68
2014	52.60	3.21	1.32	1.27	0.80

Note: *Municipal and industrial solid waste are not exclusively derived from biomass.*
Table created using source data presented in the World Bioenergy Association's Global Bioenergy Statistics (2017).

representing 87% of the total biomass supply. Most of that forest biomass was fuelwood, but also charcoal, and residual and recovered wood. Energy crops like ethanol and biodiesel only represented 3% of the total supply. For reference, biomass supply data from the World Biomass Association between 2000 and 2014 is displayed in Table 3.2, showing the dominance that solid, mostly woody, biomass has in the supply, as well as a substantial increase in its use.

Regionally, the African continent used solid biomass (wood and agriculture sources) exclusively in its primary supply of biomass—100% in 2014. Asia followed closely at over 95%. In the Americas and Europe, solid biomass represented 73% and 63% in their biomass supplies, respectively.

The continued development of bioenergy requires agricultural, forest, and other land to harvest the biomass supply. Total land area available for the biomass supply has remained nearly constant since 2000 at a little over 13,000 ha. Again, in 2014, agricultural area covered 37.6% of the total, and forests covered 30.7%. Finally, despite remaining a very small number globally, energy crops used to produce liquid biofuels have expanded tremendously in certain regions of the world. In particular, the United States and Brazil have remained the two dominant countries in terms of biofuels output (driven by government mandates and tax incentives), averaging 669,000 and 347,000 bbl/day of oil equivalent, respectively. These two countries have greatly influenced the global trend depicted in Fig. 3.28.

Global production growth is quite substantial, but when compared to the 90 million barrels of oil produced daily, as well as the small contribution to total biomass energy supply, liquid biofuels remain a small contribution to the renewable energy supply.

3.10 GEOTHERMAL SUPPLY AND PRODUCTION

Geothermal electric power plants operate under the same principles as any steam generation power plant. Three basic power plant designs that utilize hydrothermal resources include the following: (1) dry steam plants, which use steam directly to turn a turbine/generator set; (2) flash steam plants, which draw high-pressure hot water up to the surface and flash vaporizes the water to turn a turbine/generator set; and (3) binary cycle plants, which flash vaporize a secondary fluid with a

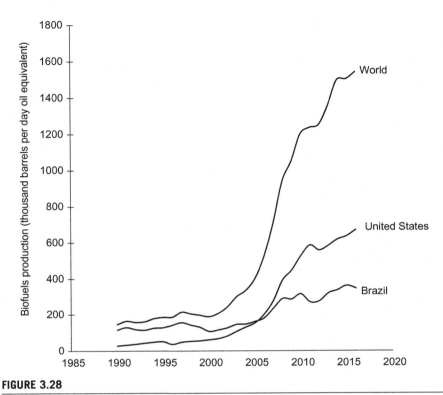

FIGURE 3.28

Global liquid biofuels production in units of barrels per day oil equivalent 1989 to 2016.

Figure created using data presented in British Petroleum's Statistical Review of World Energy (2017).

lower boiling point (e.g., butane or pentane) to also run through a turbine/genera-tor set. Condensed water brought to the surface can be reinjected into the reser-voir to generate more electricity.

The most abundant supply of highly concentrated geothermal energy exists near plate boundaries in the area known as the "ring of fire" in the Pacific Ocean. This area is where most of the current geothermal reserves are located with high concentrations of hot springs, geysers, and fumaroles. The steam or hot water can be used to generate electricity, with an estimated technical potential of 6.5 TW of thermal energy, or about 2 TW of electricity. The 2016 installed geothermal power plant capacity exceeded 13.3 GW, a 230% increase since 1990. Potential future capacity additions could see an increase to between 15 and 18 GW by 2020. Installed capacity by country is displayed in Fig. 3.29. As is expected, all these installations are within the ring of fire. Actual production in 2016 was on the order of 6.5 TWh.

Finally, we note that conduction in rocks from all geothermal activity within Earth amounts to roughly 1000 EJ per year, which is a significant potential sup-ply. This supply, in general, is readily available even outside the ring of fire; however, it requires drilling to significant depths to acquire hot enough

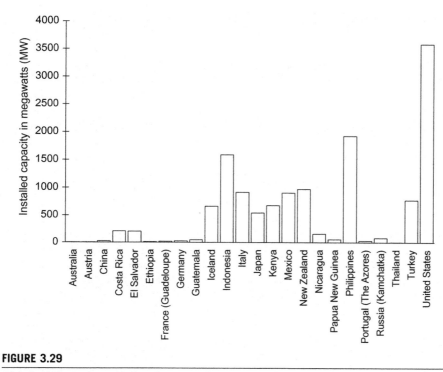

FIGURE 3.29

Global geothermal power plant installed capacity in megawatts by country 2016.

Figure created using data presented in British Petroleum's Statistical Review of World Energy (2017).

temperatures. These "hot dry rock" resources exist at depths of 5−8 km everywhere below Earth's surface. Access requires the input of cold water down an injection well, and circulation through hot fractured rock, and drawing the heated water from another well to operate a power plant. This is not currently economically feasible, though plausible in the future, since the technology required to produce this type of energy is quite similar to the methods employed in the United States to produce energy from the tight shale formations of recent years.

3.11 FUTURE ENERGY SUPPLY AND PRODUCTION

For many decades, analysts have tried to forecast both energy supply and demand. The two are mutually dependent, as we must have both energy supplies and the demand to consume them. Much of the focus has been on the ultimate depletion of finite fossil energy sources. The limits of finite sources will certainly force energy transitions—probably within the current century. In addition, concerns for environmental impacts, especially climate change, also motivate the transitions. It is commonly believed that the renewable energy sources will have little or no environmental impact. While it is almost certainly true that the non-combustion resources will have reduced impacts, it is comparably certain that some impacts will become noticeable as these alternative energy systems scale up.

The scale-up is most likely what will constrain the transition to more sustainable energy systems. The growth of solar and wind has been tremendous, and further growth will require massive manufacturing of PV cells or wind turbines. Then it will require installation, which requires suitable locations. It is reasonable to expect that there will be a saturation point of installations, at which point the growth of wind or solar power slows as fewer ideal sites remain. However, that time will likely be long from now.

A common component of all optimistic forecasts for relatively rapid transitions to more sustainable energy systems is a significant reliance on improved efficiency. Society must reduce its energy demand for there to be a realistic chance of sustainable energy systems playing a major role in the supply side. The currently popular energy sources, solar and wind, produce such a small share of the global energy supply, it will be necessary to constrain growth in energy demand to enable "new" renewable energy to gain a substantial market share in a reasonable time frame. Indeed, the savings from efficiency (and conservation) are often expected to lead to an overall decrease in global energy demand.

Whereas there is room for a great deal of net energy savings in some affluent nations, without sacrificing quality of life, it is important to remember that half of humanity lacks sufficient energy to meet basic needs. This unmet need will undoubtedly push energy demand higher, even exceeding the most optimistic estimates of efficiency and conservation measures for the 5%—10% of humanity that can potentially reduce our energy consumption to a significant degree. Efficiency measures will be essential to transitions toward more sustainable energy systems but will not be sufficient to reverse the overall growth in global energy demand until the Developing World can meet its essential needs. So, we predict that global energy demand will tend to increase over the coming decades.

If we are correct in this prediction, the physical limits on petroleum reserves are likely to limit the growth in energy production during this time frame. There is some optimism that renewable energy production will displace oil, gas, coal and nuclear power in the relatively near term; however, this optimism seems unrealistic. These four stock-limited resources, together, account for nearly 90% of global energy. Firewood, which is often not mentioned in discussing energy transitions (in spite of being the primary energy source for half of humanity and causing some 4 million deaths per year) would raise the total for resources that we seek to transition away from to over 90% of overall global energy. This goal seems unrealistic for solar and wind resources currently comprising about 2% of global energy supply—at least within the next few decades.

However, if allowed a longer time frame to transition to a more sustainable energy future, on the order of 100 years, it becomes somewhat reasonable to compare the fossil fuel reserves to likely renewable energy fluxes that can be controlled anthropogenically, probably using storage technologies in a manner that is

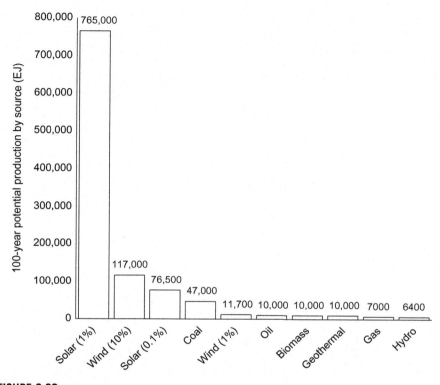

FIGURE 3.30

Comparison of possible 100-year production potentials of the renewable energy resources to current estimated proved recoverable reserves of the fossil fuels coal, oil, and gas in units of EJ.

stock- and flow based. Based on the estimated plausible renewable fluxes presented in Table 2.7 of Chapter 2, the supply and production information presented in this chapter, and thoughtful estimates of the energy needs of roughly 11 billion people by the year 2100, we present in Fig. 3.30, estimated production potentials that could be tapped in a 100-year time frame for the nonnuclear resources.

Based on these 100-year production potentials, it should be clear that solar and wind power have the most promising future potential. As shown in Fig. 3.30, if we could reliably control just 0.1% to 1% of the solar resource, we could produce between 76,500 and 765,000 EJ of energy in a 100-year time frame. Wind energy could likewise potentially produce between 11,700 and 117,000 EJ by just tapping 1%−10% of the wind resource. These numbers dwarf all but perhaps coal, which took hundreds of millions of years to convert raw biomass into fossil energy. The other renewable sources can likewise contribute to the energy mix in meaningful ways compared to the fossil fuels.

Even if annual energy consumption were to reach 1500 EJ of energy by 2100, there should be sufficient energy to meet global needs; however, it will require massive increases in renewables penetration in global energy markets, unlike what has been seen even with recent advancements. Fossil fuels will continue to be relied upon, and energy efficiency will likely be needed to stretch out the remaining reserve supply until advances in renewable energy technologies can permit this inevitable transition.

REFERENCES AND FURTHER READING

Books and Technical Articles/Reports

Blok, K., & Nieuwlaar, E. (2016). *Introduction to energy analysis*. Taylor & Francis.

Boyle, G. B. (Ed.), (2004). *Renewable energy*. Oxford University Press.

Cullen, J. M., & Allwood, J. M. (2010). The efficient use of energy: Tracing the global flow of energy from fuel to service. *Energy Policy*, *38*(1), 75–81.

Ebenhack, B. W., & Martínez, D. M. (2013). *The path to more sustainable energy systems: How do we get there from here?* Momentum Press.

Martínez, D. M., & Ebenhack, B. W. (2008). Understanding the role of energy consumption in human development through the use of saturation phenomena. *Energy Policy*, *36*(4), 1430–1435.

Martínez, D. M., & Ebenhack, B. W. (2015). *Valuing energy for global needs: A systems approach*. Momentum Press.

Meadows, D. H. (2008). *Thinking in systems: A primer*. Chelsea Green Publishing.

Prentiss, M. (2015). *Energy revolution: The physics and the promise of efficient technology*. Harvard University Press.

Scientific American. (1999). *Science desk reference*. Wiley.

Smil, V. (2008). *Energy: A beginner's guide*. Oneworld Publications.

Radovic, L. R. (1997). *Energy and fuels in society: Analysis of bills and media reports*. McGraw-Hill Custom Publishing.

Energy Information Sources and Reports

British Petroleum (BP). *BP statistical review of world energy*. (2017). Available online at: ⟨https://www.bp.com/en/global/corporate/energy-economics/statistical-review-of-world-energy.html⟩.

Energy Information Administration (EIA). *Petroleum supply monthly*. (2018). Available online at: ⟨https://www.eia.gov/petroleum/supply/monthly/⟩.

European Nuclear Society Encyclopedia. *Nuclear power plants, worldwide*. (2018). Available online at: ⟨https://www.euronuclear.org/info/encyclopedia/n/nuclear-power-plant-world-wide.htm⟩.

Geothermal Energy Association (GEA). *2016 Annual U.S. & global geothermal power production report*. (2018). Available online at: ⟨http://geo-energy.org/reports/2016/2016%20Annual%20US%20Global%20Geothermal%20Power%20Production.pdf⟩.

International Energy Agency (IEA) SolarPACES. *CSP projects around the world*. (2018). Available online at: ⟨http://www.solarpaces.org/csp-technologies/csp-projects-around-the-world/⟩.

International Hydropower Association (IHA). *2017 Hydropower status report.* (2017). Available at: ⟨https://www.hydropower.org/2017-hydropower-status-report⟩.

National Renewable Energy Laboratory (NREL). *Glossary of biomass terms.* (2011). Available at: ⟨http://www.nrel.gov/biomass/glossary⟩.

Organisation for Economic Co-operation and Development (OECD). (2016). *Uranium 2016: Resources, production, and demand.* Joint report of the Nuclear Energy Agency and the International Atomic Energy Agency No. 7301. Available online at: ⟨https://www.oecd-nea.org/ndd/pubs/2016/7301-uranium-2016.pdf⟩.

World Bioenergy Association (WBA). *WBA global bioenergy statistics 2017.* (2017). Available online at: ⟨https://worldbioenergy.org/uploads/WBA%20GBS%202017_hq.pdf⟩.

World Coal Association. *Coal electricity.* (2018). Available online at: ⟨https://www.world-coal.org/coal/uses-coal/coal-electricity⟩.

World Nuclear Association. *Information library.* (2018). Available at: ⟨http://www.world-nuclear.org/information-library.aspx⟩.

World Wind Energy Association. *2017 Statistics.* (2018). Available at: ⟨http://www.wwindea.org/2017-statistics/⟩.

Other Online Resources

Allen J. (2018). *Principles of energy conversion eBook.* Department of Mechanical Engineering—Engineering Mechanics, Michigan Technological University. Available online at: ⟨http://pages.mtu.edu/~jstallen/courses/MEEM4200/MEEM4200.html⟩.

Renewable Energy World. *Geothermal energy.* (2018). Available online at: ⟨https://www.renewableenergyworld.com/geothermal-energy/tech.html⟩.

World Energy Reference. *Energy resources: Biomass.* (2016). Available online at: ⟨https://www.worldenergy.org/data/resources/resource/biomass/⟩.

Energy–flow analyses and efficiency indicators

CHAPTER OUTLINE

In this chapter, we review the subject of energy-flow analyses by visualizing and calculating energy efficiency within the context of energy-conversion devices, processes, and macrolevel systems. Key points include the following:

- How to visualize energy flow in devices and systems?
- Energy-flow analysis in economic sectors, and
- Energy-flow efficiency indicator analysis of the 2016 US economy.

4.1 ENERGY-CONVERSION DEVICES

An energy-conversion device is simply a piece of equipment that converts one form of energy into a more useful one. Everyday examples of such devices include engines like those found in vehicles, furnaces and boilers found in residential and commercial buildings, light bulbs, and even electric power plants. They can also include devices like refrigerators, air conditioners, and heat pumps. It is easy to get bogged down describing how these devices work. For our purposes, though, it is less important to focus on the complexities of their design and operation than it is to understand the most basic functionality of those devices, which merely convert or transform the energy from one form into another form, and are governed by the laws of thermodynamics.

4.1.1 FIRST LAW

The first law of thermodynamics tells us that energy can neither be created nor destroyed, only changed, so any energy that goes into a device must (eventually) exit it. Assuming none of that energy accumulates within it, a simple energy balance of the device (or the "system") looks like

$$\text{Input energy} = \text{Output energy}$$

and is commonly illustrated as an input/output flow diagram as shown in Fig. 4.1.

If all of that converted output energy were useful, then the device's efficiency would be output/input $= 1$ or 100%. However, there are few devices where the energy conversion results in 100% useful output energy, as detailed by the second law.

4.1.2 SECOND LAW

The second law of thermodynamics deals with the disorderly process of energy conversion, in which waste is always generated during the conversion process. For example, imagine we are trying to convert the chemical energy in a fuel into mechanical work to power an engine. Many individual processes, such as incomplete combustion, friction, and conduction, can each sap some of that input energy and create heat that gets wasted or results in unburned fuel that can no longer be used. In this case, the energy balance looks something more like

$$\text{Input energy} = \text{Useful output energy} + \text{Wasted output energy}$$

and illustrated in Fig. 4.2.

Since not all of this input energy is used to complete the desired task, the device's efficiency would now be useful output/input less than 1 or less than 100%. This is the reality for most energy-conversion devices.

FIGURE 4.1

Simple first law energy input/output flow diagram. Energy is input into a conversion device and transformed into output energy.

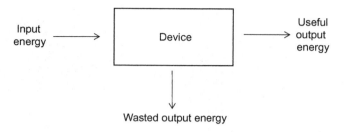

FIGURE 4.2

Simple second law input/output flow diagram. This is similar to the first law flow diagram, now accounting for unused, or wasted, energy resulting from inefficient or incomplete transformation.

Table 4.1 Efficiencies of Various Energy-Conversion Devices

Device	Transformation	Typical Efficiency or Range (%)
Photosynthesis	Radiative to chemical	1–3
Muscle	Chemical to kinetic	20
Electric heater	Electrical to thermal	100
Home gas furnace	Chemical to thermal	85–95
Incandescent bulb	Electrical to radiative	5
LED bulb	Electrical to radiative	20–50 (25% practical)
Lead acid battery	Chemical to electrical	85
Solar cell	Radiative to electrical	15–25 (practical)
Electric motor	Electrical to kinetic	90 (>200 W)
Water turbine	Kinetic to electrical	90
Gas turbine	Chemical to electrical	30–40
Automobile engine	Chemical to kinetic	20–25 (practical)

Table created using a composite of data presented in: Culp. A. W. (1991); Fowler, J.M. (1975); and Hinrichs, R.A. and Kleinbach, M. (2013).

4.1.3 COMMON ENERGY-CONVERSION DEVICE EFFICIENCIES

In Table 4.1, we list some energy-conversion devices and their estimated efficiencies of conversion. Notice that electric heaters are rated at 100% efficiency. This is one of those rare cases where the useful energy is "process heat" and the conversion from electrical energy to heat can be harnessed completely. This does not mean, however, that all of that process heat can be channeled to the desired target area. This brings about a different concept, performance, which we will describe in a later chapter. For now, think of a chicken coop that is heated at night using electric heating lamps, often just light bulbs. The coop may not be well insulated and allows some of that process heat to escape. So the efficiency of the energy-conversion device may be at or near 100%, but the performance of the system will be lower if the coop does not keep in the heat all that well.

Also notice in Table 4.1 that some of the most important energy conversions on Earth have extremely small efficiency values, such as in photosynthesis. So, although it seems desirable to value only the systems with high efficiencies of conversion, we find that reality tends to contradict this notion.

4.1.4 OVERALL SYSTEM EFFICIENCY AND LIMITING STEPS

In the previous section, we focused on just one energy-conversion device; however, there are often several devices or stages of conversion linked together that add up to the delivered energy and resulting end use that we can take advantage of. For example, the electric heater that we spoke of from Table 4.1 may indeed have a device efficiency of 100%, yet we may be more interested in knowing the overall fuel and end-use efficiency of the entire fuel-to-electricity-to-heat process. In that case, we need to determine the efficiency of conversion of the raw fuel to electrical energy at the power plant, as well as the efficiency of transporting that electrical energy to our home, as well as any other stages that lead up to our end use.

To do that, we have to isolate the relevant devices or subsystems within our overall energy conversion process (Devices 1 through N), as illustrated in Fig. 4.3.

Implicit in this more detailed picture of an energy conversion system is that multiple energy transformations may be occurring from one stage to the next, and that each transformation is generating a certain degree of wasted energy. That is, each conversion step has an inherent inefficiency, either determined by some thermodynamic limitation or by using some less than ideal technology that generates more waste than is necessary. To calculate the overall efficiency of the process we must determine the efficiency of conversion of each step and then multiply the efficiencies (or inefficiencies) of each step (1 through N) together. Expressing this as an equation, we have

$$\text{Efficiency}_{\text{Overall}} = \text{Efficiency}_{\text{Device 1}} \times \text{Efficiency}_{\text{Device 2}} \times \cdots \times \text{Efficiency}_{\text{Device}N}$$

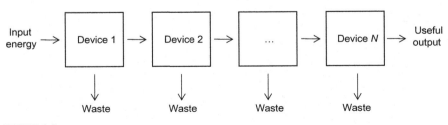

FIGURE 4.3

Multi-device, or system input/output flow diagram. A system, comprising multiple devices, can account for conversion and losses or waste at each step.

Thus, if the efficiency of the overall system is a product of the efficiencies of the individual devices/subsystems, then the degree of inefficiency of the entire process is determined by the least efficient step. This is somewhat analogous to how chemical reactions work, where a reaction rate-limiting step will affect the conversion (including rate) of initial products to final products. This is important, because it does not necessarily matter if the electric heater we mentioned earlier is 100% or 90% efficient. If one of the other devices/subsystems "upstream" of that heater is only, say, 10% efficient, then that is what will overwhelmingly define total system efficiency.

4.2 MACROSCALE ENERGY SYSTEMS AND EFFICIENCY

The same logic applied to device efficiency and system efficiency can be applied to macroscale energy systems, such as those economic sectors of a modern economy. If we isolate segments of a certain sector as "devices" of that sector, considering both supply and demand, we can visualize the energy conversion within that sector, as illustrated in Fig. 4.4.

Let us review the steps laid out in Fig. 4.4. Acquisition refers to identification, extraction, and transport of energy from primary sources (e.g., coal mining and delivery to power plants, oil and gas production, use of wind or water turbines, and harvesting raw biomass for nonenergy use). There tends to be little waste in this step. The next step, conversion, refers to the initial process involved in converting one energy form to a more useful form that devices can use (e.g., thermomechanical conversion in coal or natural gas power plants to make electricity, refining oil into many products). Waste can be rather high in this step, particularly for fossil fuel conversion. Transportation and distribution move the energy to the point of consumption (e.g., pipelines carrying natural gas to homes and businesses, the electric power grid, or moving fuel to depots and service stations). End-use conversion is the stage of finally producing useful work to the consumer via the devices we interact with (e.g., electronics, lighting, engines, and furnaces). Each step requires

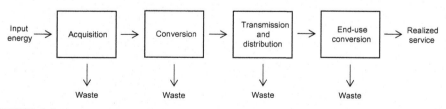

FIGURE 4.4

Energy sector input/output flow diagram. Similar to a multi-device flow diagram, individual sector energy conversion steps can be treated as individual devices to track energy transformations and losses. The concept of the sector input/output flow diagram can be further explored in Blok and Nieuwlaar (2016).

initiation from some energy input, and the final step is some realized service or product, such as warmth, lighting, transport, or a manufactured good.

Now we know that a myriad of devices perform a myriad of transformations of primary energy into fuels, electricity, heat, and other forms that get used simultaneously. We do not use only one form of energy to power all our devices to meet all our needs, but rather use a mix of primary and secondary energy to do this. At a very basic level, what we have entering our economy is already acquired primary energy flowing into established sectors of society that transfers this energy to meet needs and to achieve or maintain a high quality of life. A simplified depiction of this is illustrated in Fig. 4.5.

As we can see in Fig. 4.5, useful energy is produced from the primary energy sources (oil, gas, coal, biomass, nuclear, and renewables) and supplies the economy with all of the energy we use. A sizable mix of the primary energy supply gets diverted to the electric power sector, which then is transformed into the secondary energy carrier, electricity. Electricity is an essential element of our energy needs, but converting primary fuels into electrical energy, and then transporting it, and also having it available every second of every day can be rather inefficient, resulting in a substantial amount of waste—this is the cost of enjoying a modern economy.

Finally, it is the mix of primary and secondary energy that is supplied or delivered to the end users and is eventually transformed into other useful forms of energy through end-use technologies to provide services to consumers. Here, too, this end-use energy incurs a penalty of conversion, resulting in more waste.

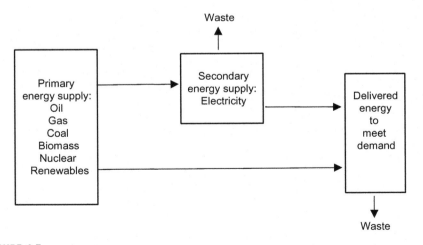

FIGURE 4.5

Simplified primary and secondary energy supply flow diagram. Primary energy flows to create secondary supply, and both primary and secondary energy becomes delivered energy for end-use demand.

4.3 ENERGY FLOWS AND EFFICIENCIES OF ECONOMIC SECTORS

All the energy we use in a system like that depicted in Fig. 4.5 is tracked by governmental agencies and regularly reported to the public as "energy consumption." We know from thermodynamics that consuming energy, in a strict physical sense, is a misnomer. As we have stated in a previous work, what gets consumed is an energy resource like coal, oil, or gas. So, when we talk about energy consumption, we are really talking about our utilization of fuels that can be ultimately converted into the many kinds of services they provide. It is not the energy itself that we want to consume. Regardless, this naming convention persists in reporting.

4.3.1 SOURCE/SECTOR LINKAGES: PRIMARY ENERGY CONSUMPTION

There are a few ways energy consumption gets reported with respect to consumption by economic sector. One of the simpler approaches used is to divide the economy into primary sectors and then track source-to-sector movement. The US Energy Information Administration (EIA) tends to separate sectors as the following: (1) electric power, (2) transportation, (3) industrial, (4) residential, and (5) commercial (also known as services in international reporting). Residential and commercial sectors consume energy in similar ways, thus these are often lumped together to simplify the picture.

This accounting method is perhaps the simplest approach to track resource use from source to sector, showing how each primary source gets divvied up into them. For example, Fig. 4.6 shows US primary energy consumption by source and sector for the year 2016 in quadrillion British thermal units (known as quads; recall a quad and an EJ are roughly the same), based on data presented by the 2017 US EIA *Monthly Energy Review*. According to the US EIA, 97.4 quads (102.8 EJ) were consumed in total in the US economy in 2016.

Fig. 4.6 is referred to as a source/sector diagram and displays two perspectives. From the "supply sources" perspective (left-hand side), Fig. 4.6 shows what share of the primary energy gets allocated or "sent" to each sector on a percent basis. From the "demand sectors" perspective (right-hand side), Fig. 4.6 also shows what share of a resource comprises a sector. With all the lines zigzagging from source to sector, we feel it is informative to look at a couple of sources and sectors individually, once from the perspective of a source/sector linkage and then from a sector/source linkage.

Let us extract the petroleum supply information from Fig. 4.6 and depict the source/sector linkage in Fig. 4.7. (Note: We are simply extracting a piece of Fig. 4.6 and presenting that extraction in Fig. 4.7, so all percentages are the same. For example, 38% of industrial sector use from petroleum in Fig. 4.6 is the same as the 38% in Fig. 4.7.)

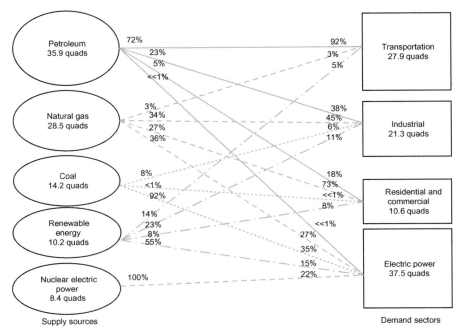

FIGURE 4.6

Source/sector-energy linkage diagram. The direction of energy flows is left to right. Percentages on the left-hand side indicate percent of the primary source flowing to each sector on the right-hand side. The percentages on the right-hand side indicate the share of that source used with respect to total energy used within that sector. Recall that quads are quadrillion Btu.

Figure created using data presented in the US EIA Monthly Energy Review (2017), following a similar graphical methodology.

What we see in Fig. 4.7 is 35.9 quads of petroleum energy were supplied to the US economy in 2016. From a supply sources perspective (left-hand side of Fig. 4.7), 72% of the 35.9 quads were directed to the transportation sector, 23% were directed to the industrial sector, 5% were directed to the residential and commercial sectors, and less than 1% were directed to the electric power sector. Conversely, from a demand sector perspective (right-hand side of Fig. 4.7), of the 27.9 quads that were consumed in 2016, 92% of the transportation sector was supplied by petroleum. Similarly, from the industrial sector perspective, of the 21.3 quads consumed, 38% of that was, again, petroleum. In the residential and commercial sectors, of the 10.6 quads consumed, 18% of that was from petroleum. Finally, in the electric power sector, of the 37.5 quads consumed, it was less than 1%.

Now, let us focus on the right-hand side of Fig. 4.6 and extract the industrial sector/source linkage, which is depicted in Fig. 4.8.

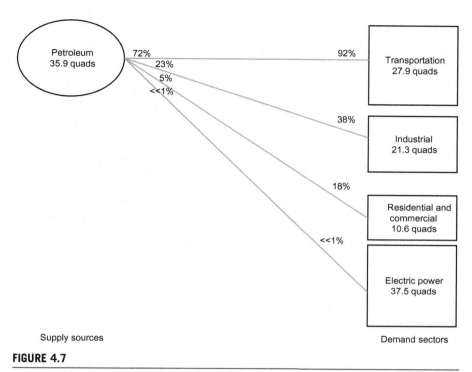

FIGURE 4.7

The petroleum supply/sector energy linkage.

Figure created using data presented in the US EIA Monthly Energy Review (2017).

What we see in Fig. 4.8 is 21.3 quads of energy were consumed in the industrial sector, and the energy mix used to satisfy these needs were supplied by petroleum (38%), natural gas (45%), coal (6%), and renewable energy (11%) that is mostly in the form of biomass. The mix is more diverse than in the transportation sector because it does not depend so exclusively on one resource. However, in total, Industry remains heavily dependent on fossil fuels at 89% total primary energy consumption. (Also note that Fig. 4.8 shows us that 23% of the petroleum supply, 34% of the natural gas supply, 8% of the coal supply, and 23% of the renewable energy supply, respectively, went to the industrial sector.)

4.3.2 DELIVERED TOTAL ENERGY BY SECTOR

Another important way to report energy consumption is to count electricity as a secondary energy supply that then feeds into the other end-use sectors, known as "delivered total energy" by sector, similar to what we depicted in Fig. 4.5. In this method of energy accounting, retail electricity sales into the demand sectors are what get reported. This is commonly how energy data is reported by the

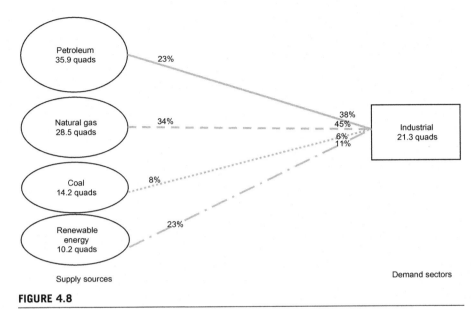

FIGURE 4.8

The Industrial sector/supply linkage.

Figure created using data presented in the US EIA Monthly Energy Review (2017).

International Energy Agency (IEA), as well as by the EIA. Depicting this added wrinkle becomes even more convoluted than before, as shown in Fig. 4.9.

Using a similar prescription for understanding the information in Fig. 4.6, what we see in Fig. 4.9 is a more accurate picture of energy end use by each sector, as well as the beginning of an accounting of the wasted energy that results from energy transformation, specifically for electricity generation. We see that of the 37.5 quads of primary energy supplied to the electric power sector, only 12.6 quads enter the other sectors as already converted electricity. Note that 97.3 quads were supplied to the economy and 72.8 quads of end-use energy were delivered, due to conversion losses.

Also of interest in reviewing Fig. 4.9 is the percentage break down of total final energy end use for each of the other energy-consuming sectors. Note that the transportation sector remains mostly unchanged between Figs. 4.6 and 4.9. In the industrial sector, we note that 13% of delivered total energy comes in the form of retail electricity; however, it is likely that more gets produced and used "in house" (e.g., the pulp and paper industry routinely produces its own electricity for utilization during their operations). Finally, note that electricity in the residential and commercial sectors represent 47% of delivered total energy.

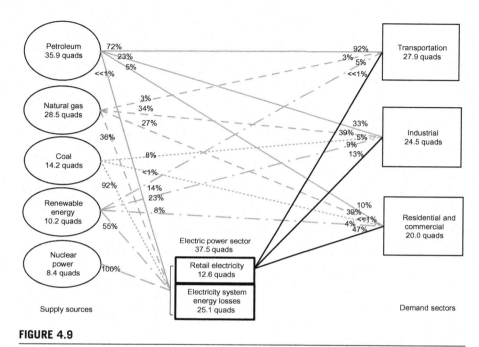

FIGURE 4.9

Delivered total energy by sector. This accounts for system losses within the electricity system.

Figure created using data presented in the US EIA Monthly Energy Review (2017), following a similar graphical methodology.

4.3.3 TOTAL ENERGY-FLOW APPROACH

A third and perhaps the most complete way to understand and visualize energy consumption is to track total energy flows using Sankey diagrams. Sankey diagrams are energy-flow diagrams that account for each individual energy input and track how these inputs flow through a system. In the United States macrolevel case, it depicts what portion of the energy supply ends up as useful "energy service" and what ends up as energy waste, or "rejected energy." A distinguishing feature is that these flow lines are line weighted to demonstrate magnitude or share of total.

The US Department of Energy-funded Lawrence Livermore National Laboratory (LLNL) produces such diagrams on an annual basis, providing analysis for how to interpret macroscale energy consumption in the United States for that year and additionally tracks energy consumption at the state level. The 2016 iteration is reproduced in Fig. 4.10. (The IEA performs a similar country-by-country analysis through their IEA *Atlas of Energy* publications on a petajoules unit basis. The reader is encouraged to review them at iea.org/Sankey.)

Looking at Fig. 4.10, we see that the LLNL significantly ratchets up the complexity in energy reporting. (In the past, we have described LLNL flow diagrams

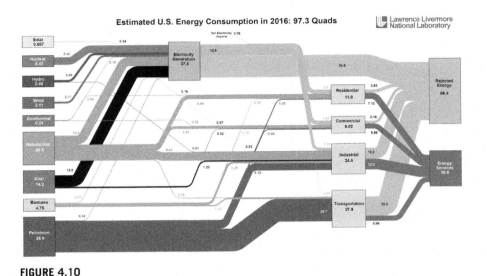

FIGURE 4.10

LLNL energy-flow (Sankey) diagram of 2016 US energy consumption.

Source: LLNL Energy Flow Charts (2017).

as Daedalian mazes, and that continues to feel like an apt description.) Basically, we see all of the 97.3 quads of energy supply separated by type on the left-hand side of Fig. 4.10, specifically solar, nuclear, hydro, wind, geothermal, natural gas, coal, biomass, and petroleum. The diagram then divvies up each of these resources into the five sector categories. Instead of percentages, each number next to each weighted line is now the quantity of energy, in quads, flowing into and out of each sector.

The LLNL chooses the same approach as described in Fig. 4.9, where primary energy first enters the electric power sector (LLNL simply calls it "electricity generation"), and retail electricity then flows into the residential, commercial, industrial, and transportation sectors. What does not enter directly into the electric power sector then gets divvied up into the remaining sectors from the primary supply. On the right-hand side of Fig. 4.10, the LLNL reports that of the 97.3 quads that entered the economy, only 30.8 quads or 31.6% of the total input resulted in useful energy services, with the rest as wasted energy, or "rejected energy," as labeled in the figure. Thus, the US economy's simple energy efficiency is 30.8/97.3 = 0.316 or 31.6%.

4.3.4 SECTOR ENERGY FLOWS AND SIMPLE EFFICIENCY ANALYSIS USING THE SANKEY APPROACH

Choosing and then gleaning appropriate information from any of the three methods of energy accounting depends on what the analyst is looking for. Obviously, if it is to understand total primary consumption information, then the first method depicted by Fig. 4.6 will suffice and that is typically adequate for understanding sector mixes, which may drive policy changes. For example, having petroleum represent 92% of the entire transportation sector maybe a negative choice if you are concerned about energy security and you are unable to control the petroleum supply. Alternatively, if the analyst's goal is to understand total delivered energy, the methodology depicted in Fig. 4.8 will suffice. Lastly, if it is important to do scenario building and forecasting with an eye for supply and efficiency of conversion, then the Sankey approach in Fig. 4.10 likely is the most informative.

Note that the Sankey approach tracks both primary energy consumption and total delivered energy simultaneously, thus residential, commercial, and industrial sector efficiency analyses can get complicated by this feature, because calculating the efficiency of conversion could be different if it is evaluated on a total basis versus an end-use basis. Most reporting agencies, however, focus on end-use conversion efficiency because most of the rejected energy at that point would likely be due to direct combustion of fuels being delivered for heating/cooling and transport. (There is a simple way to account for this, which will be described later in this chapter.)

4.3.4.1 Electricity sector energy-flow analysis

According to the estimated LLNL data reported in Fig. 4.10, in 2016, the electric power sector consumed 37.5 quads of total energy supply, with shares presented graphically in Fig. 4.11. (Small rounding errors exist when adding percentage totals. Two decimal places displayed for the benefit of the smaller contributors.)

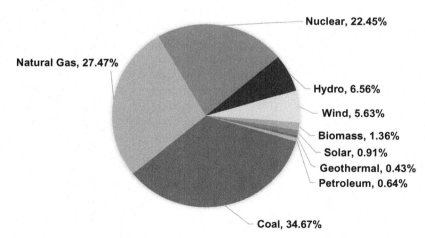

FIGURE 4.11

US electric power energy supply by share of source (total: 37.5 quads).

Figure created using data presented in the 2016 US Energy Consumption LLNL Energy Flow Chart (2017).

The electric power sector is by far the largest consumer of the US energy supply, consuming 37.5 quads of primary energy. The 2016 energy mix was diverse, with a little less than 35.0% (13 quads) coming from coal, about 27.5% (10.3 quads) coming from natural gas, 22.5% (8.42 quads) coming from nuclear power, and about 15.0% (5.6 quads) coming from renewable energy sources. In recent years, due in large part to hydraulic fracturing/horizontal-drilling techniques, and to a large increase in wind farm installations, the US has seen coal's share decrease substantially, while natural gas and renewable energy have taken that share. Total consumption has fallen from the 2007 US high of 101.5 quads, which has also affected source contributions, and thus percentages.

Also, according to the LLNL estimates in Fig. 4.10, the Electric Power sector converted 37.5 quads of primary energy into 12.6 quads of useful secondary energy (electricity) that fed into the other end-use sectors, thus wasting 24.9 quads as unused or rejected energy. This means that the sector's simple primary energy conversion efficiency was 12.6/37.5 = 0.336 or 33.6%. Thus, 24.9 quads or 66.4% of primary energy supply fed into that sector was wasted immediately. In practice, this value represents an average of the chief energy efficiency indicator for the electric power sector: fuel conversion efficiency. Conversion efficiencies of all nonrenewable electric energy generators are tabulated in Table 4.2 (Renewable energy conversion inefficiencies for wind, hydro, and solar photovoltaic (PV) are not counted since the unit of fuel, be it wind, water, or sunlight, originates from a nominally nondepletable, noncombustible, nonretail source. 1 quad of electricity produced equals 1 quad of primary energy supplied.)

Table 4.2 Typical Fuel Conversion Efficiencies of Electric Energy Generators

Device	Transformation	Typical Efficiency (%)
Traditional hydro	Kinetic to electrical	90
Natural gas (combined cycle)	Chemical to electrical	45
Natural gas (internal combustion)	Chemical to electrical	37
Petroleum (combined cycle)	Chemical to electrical	35
Coal (steam generation)	Chemical to electrical	34
Petroleum (steam generation)	Chemical to electrical	33
Natural gas (steam generation)	Chemical to electrical	33
Petroleum (internal combustion)	Chemical to electrical	33
Nuclear (steam generation)	Nuclear to electrical	33
Natural gas (gas turbine)	Chemical to electrical	30
Wind	Kinetic to electrical	26
Petroleum (gas turbine)	Chemical to electrical	25
Solar thermal (STEC)	Thermal to electrical	21
Geothermal	Thermal to electrical	16
Solar PV	Radiative to electrical	12

Table created using a composite of data presented in: US EIA Electric Power Annual (2016), and US EIA Annual Energy Review: Conversion Efficiencies of Noncombustible Renewable Energy Sources (2010).

Conversion efficiencies for thermal power plants (fossil fuel and nuclear power) listed in Table 4.2 are determined by the power plant's reported heat rate. Heat rates are simply expressions of how much thermal energy must be expended to obtain a unit of useful work, in this case, a 1 kW unit of electricity. The US reports heat rates as Btu/kWh, or how many Btu/hour of energy are required to produce 1 kW of work. To determine the efficiencies of thermal power plants, then, one must divide 3412 by reported heat rates (because 3412 Btu is roughly equal to 1 kWh). For example, the heat rate of a typical coal-fired powered plant is about 10,000 Btu/kWh, giving a conversion efficiency of about 34%. For comparison, coal-fired power plant energy conversion efficiency in the year 1900 was closer to 17%, and the ideal thermodynamic efficiency (i.e., Carnot efficiency) operating at typical power plant temperatures is roughly between 60% and 65%.

Despite the doubling or more in efficiency since 1900, there does remain room for improvement, and, indeed, there are a number of companies trying to improve efficiencies in many of these systems. Most notable are natural gas turbine systems, which have efficiencies now exceeding 50%. The caveat is that this depends on the way the technology is used. Natural gas turbines often operate as peak electricity power generators (to be described in a later chapter), causing reduced optimal system efficiency. Stringent pollution controls or change in fuel type or grade can also affect heating rates, thus causing decreased thermal efficiency for the power plant.

4.3.4.2 Transportation sector energy-flow analysis

According to the estimated LLNL data reported in Fig. 4.10, in 2016, the transportation sector consumed 27.9 quads of total delivered energy, with shares presented graphically in Fig. 4.12.

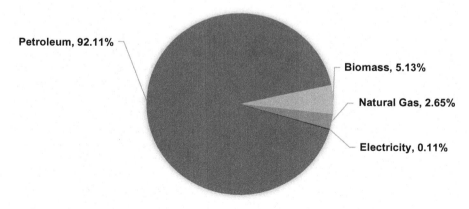

FIGURE 4.12

US transportation delivered total energy by share of source (total: 27.9 quads).

Figure created using data presented in the 2016 US Energy Consumption LLNL Energy Flow Chart (2017).

Transportation is the second largest consumer of energy in the US system, consuming 27.9 quads. Although there is some secondary energy in the supply mix, energy consumption in the transportation sector is almost exclusively primary energy, and the overwhelming majority comes from petroleum (in the forms of motor gasoline, diesel, jet fuel, and residual fuel oil). The sector remains dominated by the petroleum resource and by extension the internal combustion engine. Biomass has increased its share in the sector in recent years due to a concerted effort to expand the use of fuel alcohol (the majority corn ethanol) and biodiesel. Natural gas has also slightly increased as more supply has come online in recent years. A paltry amount of the electricity supply is used in transportation, primarily for rail. Despite a large desire to transition to fully electric vehicles in recent years, petroleum will likely continue to dominate the sector for the foreseeable future.

Also, according to the LLNL Sankey diagram, the transportation sector converted 27.9 quads of primary and secondary energy into 5.86 quads of energy service and wasted 22.0 quads as unused or rejected energy. This means that the sector's simple total delivered end-use energy efficiency was 5.86/27.9 = 0.21 or 21%. In practice, this represents the conversion efficiency of all passengers and freight transport combined, which includes

- light duty vehicles,
- heavy duty vehicles,
- powered 2- and 3-wheelers,
- buses,
- trains,
- planes, and
- ships.

Total fuel conversion efficiency in transportation comprises two components: engine efficiency and mechanical efficiency. Engine efficiency is the same as thermal efficiency and deals with how much chemical energy in the fuel (gasoline, diesel, etc.) is converted into work needed to move the pistons in the engine—like how electric power plants turn turbines. Mechanical efficiency is the portion of the work delivered by the engine to the drivetrain, moving objects along. Thermal efficiency for standard gasoline engines can range from 20% up to 38%, with an average of roughly 25%, while Diesel engines average closer to 35%. Mechanical efficiency at cruising speeds is roughly 50%, due to transmission loss, air drag, rolling resistance of the tires, and friction within the engine. Obviously, engine size, speed, and cargo load (either passenger number or size of freight) all impact these numbers.

The bulk of transport in all regions of the world, and especially in the US, depends on the automobile. The primary metric used to assess energy efficiency in the automotive segment of the transportation sector is generally fuel economy. In the US, this is often reported as miles per gallon of gasoline equivalent and in most other regions is reported as liters gasoline equivalent per 100 km. While equivalent, the latter is often more useful for doing energy analyses. When including other modes of transport, it also is useful to include unit person or cargo weight bases, to give a more complete estimate of the fuel economy per person or per freight moved by that mode of transport.

Finally, it is important to keep in mind that the estimated rejected energy can change based on how fuel efficiency is either measured or estimated by a regulatory body. For example, over the past decade, the US Environmental Protection Agency (EPA) has been altering the method by which fuel economy is estimated, using more realistic "real world" testing protocols. This has resulted in a decline in the end-use energy efficiency from about 25% as recently as 2010 to the current estimate of 21%. (Note: transportation end-use efficiency had been estimated to be roughly 25% going back to 1950 as indicated in the LLNL's first report in 1973.) Also keep in mind that the US corporate annual fuel economy standards that are followed by all automobile manufacturers and that get reported to the public are often 20% higher than these used here, which are adjusted, real-world values.

4.3.4.3 Industrial sector energy-flow analysis

According to the estimated LLNL data reported in Fig. 4.10, in 2016, the industrial sector consumed 24.5 quads of total delivered energy, with shares presented graphically in Fig. 4.13.

The industrial sector is the third largest consumer of the US energy supply, consuming 21.31 quads of primary energy and 3.19 quads of electricity, for a total of 24.5 quads. Energy sources are commonly used in the industrial sector as boiler

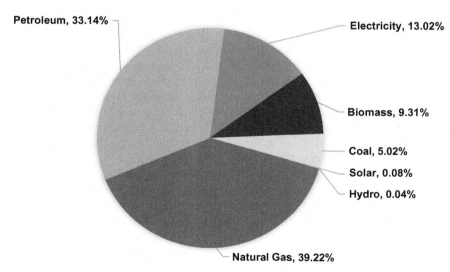

FIGURE 4.13

US industrial delivered total energy by share of source (total: 24.5 quads).

Figure created using data presented in the 2016 US Energy Consumption LLNL Energy Flow Chart (2017).

fuel to generate steam, hot water, and electricity, as well as for process heat to raise the temperature of products in manufacturing. In addition, a rather large percentage of the energy sources used in this sector is used as feedstocks or raw materials to make manufactured goods. These feedstocks include liquid petroleum gas (LPG), coal, natural gas, and other less common sources derived primarily from biomass, including wood and wood residues, agricultural waste, and others. All feedstocks can be consumed in other sectors, so they get treated in reporting as if they were used for energy consumption. Manufacturing facilities also use natural gas and electricity for heating, cooling, lighting, and for powering equipment and appliances, as would be used in the residential and commercial sectors that we will describe later.

Also, according to Fig. 4.13, the industrial sector utilized 12.0 quads of end-use energy as energy service and wasted 12.5 quads as unused or rejected energy, meaning the sector's simple total delivered end-use energy efficiency was 12.0/24.5 = 0.49 or 49%. Although industry comprises several different enterprises, the largest consumers of energy in this sector are

- petroleum refining,
- chemical manufacturing,
- pulp and paper manufacturing, and
- metals mining and manufacturing.

Combined, these four industries consume practically all the energy feedstocks and over half the energy consumed for heat, power, and electricity generation. Significant energy losses are to be expected in these industries, especially in manufacturing processes that require extremely high operating temperatures. It also provides incentives to recycle already processed materials and integrate them into manufacturing as much as possible. (Note: relatively smaller industries include food and agricultural goods, cement, textiles, and heavy equipment manufacturing, among others. Their contributions sometimes can be much larger in different postindustrial regions or in regions with mega populations.)

Total conversion efficiencies in the industrial sector are determined by energy used per value added to the economy. This type of efficiency is slightly different than the type of efficiency that we have described thus far, where the efficiency indicator or metric is now equal to input energy divided by the quantity of manufactured good, as depicted in Fig. 4.14.

For petroleum refining, this is energy input in joules per liter or per kilogram of refined fuel. Similarly, for chemicals manufacturing, it is energy input per kilogram of industrial chemical produced. For aluminum manufacturing, it is energy input per kilogram of manufactured aluminum good produced. All of these efficiencies, then, are controlled by the total energy needs for producing that valued good and improving the efficiency depends on how to produce an equal amount of that good for less total energy expended.

Finally, note that prior to 2016, the end-use efficiency of the industrial sector had always been estimated to be close to 80% as opposed to the 49% that is now estimated. This noticeable change is the result of a different and presumably more

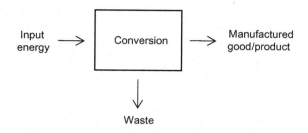

FIGURE 4.14

Input/output diagram for a manufactured good. Similar to a second law flow diagram, however, the output is a physical good.

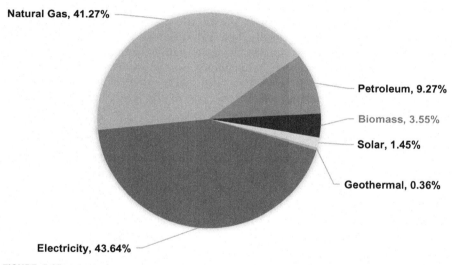

FIGURE 4.15

US Residential delivered total energy by share of source (total: 11 quads).

Figure created using data presented in the 2016 US Energy Consumption LLNL Energy Flow Chart (2017).

accurate methodology employed by the US DOE to analyze manufacturing energy consumption. As with the US EPA and the transportation sector, more accurate accounting methods can change the final sector end-use efficiency calculation, often reducing the traditionally reported value.

4.3.4.4 Residential sector energy-flow analysis

According to the estimated LLNL data reported in Fig. 4.10, in 2016, the residential sector consumed 11 quads of total delivered energy, with shares presented graphically in Fig. 4.15.

The residential sector is the fourth largest consumer of the US energy supply, consuming 6.2 quads of primary energy and 4.8 quads of electricity, for a total of 11 quads. Of the 6.2 quads of primary energy, most—just under 90%—comes from natural gas and petroleum products. The remainder comes from biomass, primarily wood, as well as solar and geothermal. Natural gas and petroleum (heating oil and LPG) are used primarily in boilers and furnaces to generate steam, hot water, or forced hot air to heat residential buildings (single and multifamily units) and for washing and bathing. Biomass is used in stoves for distributed space heating or in centralized boiler/hot water heating systems. Electricity is used for not only cooling/air conditioning but is also used for heating in certain regions. Electricity is also heavily used to light the inside and outside of buildings and for powering appliances and small devices. Of course, natural gas, LPG, and electricity are all used for cooking as well.

Note that the 4.8 quads of electricity used in the residential sector represents 38% of all the electricity produced by the electric power sector. This is a rather large portion of the electricity supply, and it is beneficial to understand the magnitude of the primary energy supply impact of using electricity. A simple way others have reported to estimate the combined supply impact of using both primary energy and electricity is to add estimated fuel use to electricity use, accounting for conversion efficiency in the production of that electricity. This can be expressed as

$$\text{Primary energy use} = \text{fuel use} + \text{electricity use/conversion efficiency}$$

In the case of the residential sector, total primary energy used would then be estimated as 6.2 quads + 4.8 quads/0.336 = 20.49 quads of primary energy. This assumes the conversion efficiency of 33.6%, as indicated by the simple efficiency calculated for the electric power sector described earlier.

Electricity is a clean and robust form of energy at point of use, so efficiency loss is obviously an accepted cost of using electricity (because it extends the day, offers mobility, reduces local air pollution, etc.); however, it does point to the potential for developing more efficient electrical energy conversion systems or for using more renewable energy supply, including rooftop renewable energy systems, to reduce nonrenewable supply needs and the associated pollution that results from the combustion of fossil and biomass fuels.

This wrinkle of needing to account for electrical energy conversion is what the LLNL analysis of Fig. 4.10 avoids, by separating the electric power sector out from the end-use sectors and feeding just the delivered, retail electricity. Ultimately, end users are most directly impacted by the energy they use in the form that is delivered to them, so this is a logical approach. Thus, according to Fig. 4.10, the residential sector utilized 7.12 quads of end-use energy as energy service and wasted 3.83 quads as unused or rejected energy, meaning the sector's simple total delivered end-use energy efficiency was 7.12/11 = 0.65% or 65%. In practice, this represents the combined conversion efficiencies of all the end-use energy systems used to perform the following desired tasks: (1) heating, including water heating; (2) cooling; (3) lighting; and (4) powering appliances and devices, including for cooking.

Each of the systems used to accomplish the above tasks has different conversion efficiencies and is impacted by the overall performance of the different systems in tandem with the building's characteristics, and with the local environmental conditions (particularly climate). They are also impacted by individual end-user choices, personal comfort levels, and the willingness or ability to pay for services. Ultimately, what we seek from these end uses is to achieve comfort in our homes and to (ideally) realize that comfort using the smallest amount of energy.

As such, the conversion efficiency of residential heating and cooling systems depends on the thermal efficiency of the heating systems, the energy distribution efficiency, and the building's overall ability to perform its function in keeping a desired ambient temperature. Likewise, lighting conversion efficiency depends on the lighting technology being employed, which often depends on function and building design. Appliances and devices used for all other household needs have their own conversion efficiencies as well. These, of course, are complicated by the age of each system.

Thus, determining efficiency becomes extremely site-specific; however, according to Blok and Nieuwlaar, general metrics or indicators that can be used in determining end-use residential sector efficiency include the following:

- For space heating and cooling: Temperature-corrected energy per floor area,
- For water heating and cooking: Energy per occupied dwelling,
- For lighting: Luminous energy per floor area, and
- For appliances: Energy per appliance unit.

These are general population activities structured as floor area or dwelling per population, thus, the primary reporting units in residential energy sector tracking is energy used per occupant or household.

4.3.4.5 Commercial sector energy-flow analysis

According to the estimated LLNL data reported in Fig. 4.10, in 2016, the commercial sector consumed 9.02 quads of total delivered energy, with shares presented graphically in Fig. 4.16.

The commercial sector is the smallest consumer of the US energy supply, consuming 4.38 quads of primary energy and 4.64 quads of electricity, for a total of 9.02 quads. Of the 4.38 quads of primary energy, 94% comes from natural gas and petroleum. The remainder comes from biomass, primarily wood, as well as solar, geothermal, and coal. Note that the 4.64 quads of electricity used in the commercial sector represent 36.8% of all the electricity produced by the electric power sector. Looking at the data in Figs. 4.10 and 4.16, it should be clear why the commercial and residential sectors are often lumped together, as the two sectors consume energy in almost identical ways.

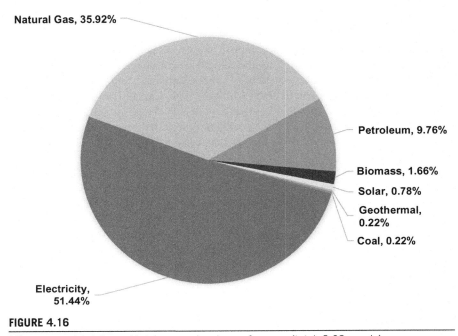

FIGURE 4.16

US commercial delivered total energy by share of source (total: 9.02 quads).

Figure created using data presented in the 2016 US Energy Consumption LLNL Energy Flow Chart (2017).

The commercial sector consists of places of business, service and care, and community including offices and banks, schools, warehouses, hotels, hospitals, shopping centers, and places of worship. As in the residential sector, the structural units used to analyze energy use in the commercial sector are buildings; however, these buildings are much larger in size and their main activities are value-added services. For example, the function of energy use in this sector is not just to provide heat to a building for people but rather to provide heat to a building to provide a comfortable work environment for employees to carry out service activities that result in added value to the economy or community.

This value-added function of the commercial sector means you are still providing heating and cooling, lighting, and power for appliances for use by people in ways very similar to the residential sector, but now the aim of that end use is often financial profit for the provider and a physical or emotional value from a service that consumers want or need. Examples can range from grocery stores to restaurants, to insurance products, to retirement investments. In these instances, the most measurable efficiency indicator would be energy use per unit area or volume of office space and/or energy use per office equipment unit. However, the more challenging element is, the more complicated it is to determine the real value added to society for the service being offered and how energy consumption can most efficiently be utilized to do that.

Finally, according to Fig. 4.10, the residential sector effectively utilized 5.86 quads of end-use energy as energy service and wasted 3.16 quads as unused or rejected energy, meaning the sector's simple total delivered end-use energy efficiency was 5.86/9.02 = 0.65 or 65%. This efficiency is the same as in the residential sector. Again, in practice, this represents the combined conversion efficiencies of all the end-use energy systems used to perform the following desired tasks: (1) heating, including water heating, (2) cooling, (3) lighting, and (4) powering appliances and devices, including for cooking. Of these, space heating, lighting, and cooling are the three largest end uses.

4.3.4.6 Side-by-side sector comparisons

Based on the information provided in the LLNL Sankey diagram, we present side-by-side graphical comparisons of energy transformation (used versus wasted) of the five sectors, displayed in Fig. 4.17.

As we have used the US energy system as a microcosm of the world energy system, there is some "take home" information that we should be mindful of when looking at macroscale energy data as a whole. The primary ones are detailed below.

First, in the electric power sector, it takes approximately 3 units of primary energy to produce 1 unit of useful end-use energy (take the gray bar of "used energy" in the left of Fig. 4.17 and imagine stacking three of them to get the darker bar of total energy). Put another way, for every 100 units of primary energy used to create electricity, 34 units are useful electrical energy and 66 units are waste. If we continue to use fossil fuel (and nuclear powered) devices to produce

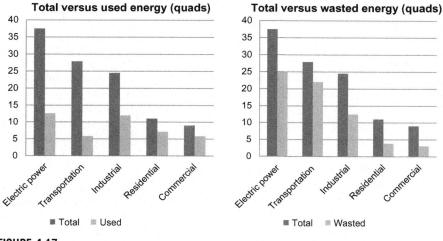

FIGURE 4.17

"Total" versus "used" and "total" versus "wasted" energy in the US energy system.

Figure created using data presented in the 2016 US Energy Consumption LLNL Energy Flow Chart (2017).

our electricity, there remain opportunities to improve this ratio; however, based on the laws of thermodynamics, our maximum possible conversion is roughly 65 units of energy service and 35 units of waste.

Second, in the transportation sector, it takes approximately 5 units of mostly primary energy to produce 1 unit of useful transport energy. In this case, for every 100 units of primary energy used to transport people and goods, 21 units actually do so, while 79 units are wasted. As mentioned earlier, efficiency losses are not just thermal but also mechanical. It is possible to improve automobile efficiencies through design, both in the power train and in the physical design of the vehicle. This plausibly could improve sector efficiency enough so that 40−50 units are useful, while 50−60 get rejected. It maybe more beneficial, however, to invest in infrastructure to maximize per person and per cargo efficiency, such as in mass transit. The United States can look to successes in other regions of the world for improving efficiency in this sector.

Third, in the industrial sector, it takes approximately 2 units of primary and secondary energy to produce 1 unit of service or good. The industrial sector is peculiar because it uses so much primary energy as feedstock and produces a significant portion of its own thermal energy and electricity for producing manufactured goods. In addition, the bulk of consumption in this sector is concentrated on four primary subsectors, which all are rather energy intensive. However, these companies that comprise the sector tend to be publicly traded, so there is probably the largest incentive in this sector to be as energy efficient as possible to maximize profits for themselves and for their respective stakeholders. The most effective strategy employed by this sector involves recycling of materials to reduce the mining and thermal needs of manufacturing goods from virgin stocks, as well as using waste by-products for generating process heat and electricity.

Fourth, in the residential and commercial sectors, it takes approximately 1.5 units of primary and secondary energy to produce 1 unit of end-use energy for keeping comfortable and for value-added services. While much of the energy use in these sectors is to meet basic human needs, there also is a great deal of discretionary energy use that wastes energy. Also, the primary consumption unit is the building, thus it is important to assess the efficiency of energy-conversion devices within the buildings that deliver heat, light, and other utilities, as well as to assess the performance of the building stock at minimizing the specific energy use per unit area or per person or household. Even at a nominal 65% end-use efficiency, there remain an abundance of opportunities to improve energy use in these sectors.

Finally, the industrial, residential, and commercial sectors all use a substantial amount of electricity. By parsing out electric power as its own sector, we obscure the true primary energy impact of these sectors. So, while it may be useful to calculate end-use efficiency for the purposes of the LLNL analysis, which may be done to ultimately look at the efficiency of end-use technologies (e.g., refrigerators or lighting systems), it is also important to look at what the primary energy efficiency of conversion is.

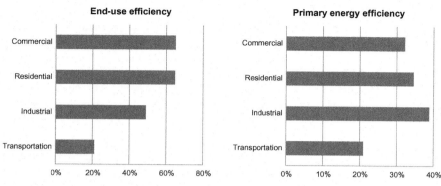

FIGURE 4.18

Side by side comparison of end use versus primary energy efficiency of the main energy use sectors of the US economy.

Figure created using data presented in the 2016 US Energy Consumption LLNL Energy Flow Chart (2017).

By adding fuel use to electricity use, and dividing the latter by the efficiency of electricity conversion, we can embed the electric power sector into the other sectors to gauge their total primary energy-conversion efficiency. A side-by-side comparison is provided graphically in Fig. 4.18. What we see is a much more uniform efficiency among the sectors (industrial 39%, residential 35%, and commercial 32%), although transportation still lags (21%). It also shows that improving the efficiency of electricity conversion, by default, improves the efficiency of the other sectors. It also shows us that because we want electricity available to us at all hours of the day, we have embedded waste that will be challenging to shed using just technology.

4.4 SUMMARY

This chapter has introduced and reinforced several concepts for characterizing energy flows and efficiency. These include energy input/output diagrams and specifically use an input/output diagram approach to track energy flows through energy systems. Inputs are often in the form of fuels derived from primary-energy supplies. Outputs maybe more useful energy forms used to perform necessary and wanted tasks but also can be goods that require material and energy transformations in their manufacture.

Regardless of how many conversion steps are involved in going from primary energy source to end-use work, there are efficiency losses at every step. Simple input/output models can be used to track conversion efficiencies of devices and systems of devices, ranging from photosynthesis to national and global economies. The detail with which an energy analysis can proceed depends on the goals of the analysis. When applied to macroscale energy systems, energy and energy efficiency analyses can lead both to broad and detailed understanding of energy as it flows through devices and systems.

Using the US energy system as a macrolevel case, the reader is expected to now understand the energy conversions that take place in common energy-consumption sectors of modern economies. This knowledge will be helpful for understanding systems, processes, and devices used on a daily basis.

REFERENCES AND FURTHER READING

Books and Technical Articles/Reports

Aubrecht, G. J. (2006). *Energy: Physical, environmental, and social impact* (3rd ed.). Pearson Prentice Hall.

Blok, K., & Nieuwlaar, E. (2016). *Introduction to energy analysis.* Taylor & Francis.

Culp, A. W. (1991). *Principles of energy conversion.* McGraw Hill.

Ebenhack, B. W., & Martínez, D. M. (2013). *The path to more sustainable energy systems.* Momentum Press.

Fowler, J. M. (1975). *Energy-environment source book.* Washington: National Science Teachers Association, Energy-Environment Materials Project.

Hinrichs, R. A., & Kleinbach, M. (2013). *Energy: Its use and the environment* (5th ed.). Brooks/Cole Cengage.

Radovic, L. R. (1997). *Energy and fuels in society: Analysis of bills and media reports.* McGraw-Hill Custom Publishing.

Reynolds, J. P., Jeris, J., & Theodore, L. (2007). *Handbook of chemical and environmental engineering calculations* (2nd ed.). Wiley Interscience.

Schroeder, D. V. (2000). *An introduction to thermal physics.* Addison Wesley Longman.

Energy Information Sources and Reports

Energy Information Administration (EIA). *Annual energy review: Conversion efficiencies of noncombustible renewable energy sources.* (2010). Available from <https://www.eia.gov/totalenergy/data/annual/pdf/sec17.pdf>.

Energy Information Administration (EIA). *Monthly Energy Review.* (2017). Available from <https://www.eia.gov/totalenergy/data/monthly/index.php>.

Energy Information Administration (EIA). *Electric power annual: Average tested heat rates by prime mover and energy source, 2007−2015.* (2016). Available from <https://www.eia.gov/electricity/annual/html/epa_08_02.html>.

Energy Information Administration (EIA). *Commercial buildings energy survey.* (2017). Available from <https://www.eia.gov/consumption/commercial/>.

Energy Information Administration (EIA). *Manufacturing energy consumption survey.* (2017). Available from <https://www.eia.gov/consumption/manufacturing/>.

Energy Information Administration (EIA). *Monthly energy review.* (2017). Available from <https://www.eia.gov/totalenergy/data/monthly/>.

Energy Information Administration (EIA). *Residential energy consumption survey.* (2017). Available from <https://www.data.gov/manufacturing/manufacturing-energy-consumption-survey/>.

International Energy Agency (IEA). *Atlas of energy: Energy balance flows.* (2016). Available from <https://www.iea.org/Sankey/>.

International Energy Agency (IEA). *Energy efficiency indicators highlights.* (2016). Available from <https://www.iea.org/publications/freepublications/publication/energy-efficiency-indicators-highlights-2016.html>.

Lawrence Livermore National Laboratory (LLNL). (2017). *Estimated U.S. energy consumption in 2016.* Lawrence Livermore National Laboratory (LLNL) Report LLNL-MI-410527. Available from <https://flowcharts.llnl.gov/commodities/energy>.

Other Online Resources and Data Sources

Allen, J. (2018). *Principles of energy conversion ebook.* Department of Mechanical Engineering — Engineering Mechanics. Michigan Technological University. Available from: http://pages.mtu.edu/~jstallen/courses/MEEM4200/MEEM4200.html.

Nowling, U. (2015). Understanding coal power plant heat rate and efficiency. *Power Magazine.* Powermag.com. Available from <http://www.powermag.com/understanding-coal-power-plant-heat-rate-and-efficiency/>.

Pisupati, S. V. (2017). *John A. Dutton e-Education Institute course on energy conservation and environmental protection.* Pennsylvania State College of Earth and Mineral Sciences. Available from <https://www.e-education.psu.edu/egee102/l1.html>.

Electric power sector energy efficiency

5

CHAPTER OUTLINE

In this chapter, we provide descriptions of electricity generation and delivery, including concepts and calculations related to:

- customer demand profiles,
- efficiencies of conversion and electrical transmission.

5.1 GLOBAL ELECTRICITY CONSUMPTION: BASIC STATISTICS

The world's nations are consuming around 25,000 TWh of generated electricity annually from a total installed generating capacity of over 15 TW. For context,

Energy Efficiency. DOI: https://doi.org/10.1016/B978-0-12-812111-5.00005-6

China consumes around 5000 TWh; the United States around 4000 TWh; Germany and Brazil around 500 TWh each; the UK around 300 TWh; and Nigeria around 24 TWh. On a per capita basis, poorer nations with significant populations (like Nigeria) can consume as little as 130 kWh per person per year, while wealthy, highly developed nations can consume well over 10,000 kWh per person. (Recall that 1 TWh equals 1 billion kWh.)

Since electricity is a hallmark of an advanced economy, we can expect more nations to invest in more electrical infrastructure and even more generating capacity. Indeed, global electricity generation has risen from 15,000 to 25,000 TWh in just the past 15 years and growth is expected to continue at a rapid pace, as developing countries continue to incorporate electrical utilities into their respective infrastructures. Consider what it would mean if the entire world were to reach the consumption levels of highly developed nations, 10,000 kWh per person, with 7.6 billion people on the planet.

$$7.6 \times 10^9 \text{people} \times 10,000 \text{ kWh/person} = 76 \times 10^{12} \text{ kWh} = 76,000 \text{ TWh}$$

It is more than three times the global energy consumption at the time of the writing of this book. Of course, if the population doubles, which it is on pace to do before the end of the 21st century, it would require six times the current electric power generation to meet this target. Efficiency will be absolutely essential in order to rein in electricity demands as population continues to grow and demand for electricity to support development grows.

As was the case in our analysis of the United States energy system in Chapter 4, Energy Flow Analyses and Efficiency Indicators, fossil and nuclear fuels supply the bulk of total electric power globally. According to International Energy Agency (IEA) data, combined, these nonrenewable fuels comprise over 70% of global electricity generation. Fig. 5.1 depicts electric power source shares from 2015 IEA reporting, and we see that coal and natural gas, specifically, are

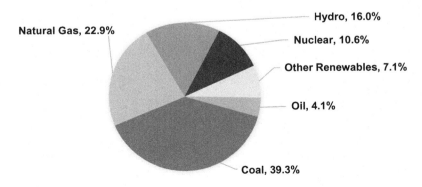

FIGURE 5.1

2015 World electric power supply by share of source (total: 24,255 TWh).

Figure created using data presented in IEA Key World Energy Statistics 2017 (2018).

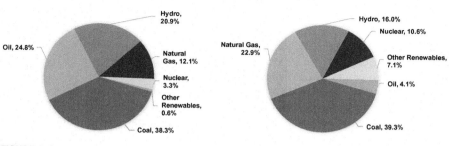

FIGURE 5.2

1973 and 2015 World electric power supply by share of source (1973 Total: 6131 TWh; 2015 Total: 24,255 TWh).

Figure created using data presented in IEA Key World Energy Statistics 2017 (2018).

the most significant in the electrical power supply, with hydropower also being a major contributor. (*Note*: Hydro is often separated from other renewable sources in IEA reporting, which include wind, geothermal, solar, biofuels, etc., as historically it tended to dwarf the share from these small but quickly growing sources.)

In addition, the IEA always tends to reference most recent consumption to 1973 (IEA was established in 1974, in part, because of the global oil crisis of 1973) and if we compare historical supply by share of source, as in Fig. 5.2, the most significant difference is a major reduction in oil use for electricity generation—24.8% of the total share in 1973 versus 4.1% in 2015. In its place grew shares of natural gas, nuclear, and non-hydro renewables. Coal slightly increased in share specifically as nuclear power decreased in recent years. (For example, Japan and Germany have significantly reduced nuclear power in favor of coal and renewables.)

The total difference in terms of total shares of electricity generated from non-renewable energy is about 5.5% (a decrease from 78.5% in 1973 to 73% in 2015). There are many reasons why the use of nonrenewable energy persists at the levels they do in the electricity supply, and many good reasons to improve the efficiency of conversion and/or expand the use of renewable energy in the electricity supply going forward.

On the one hand, fossil and nuclear fuels are highly transportable and extremely energy dense. This allows both for steady, baseline electricity that is available at all hours, as well as electricity that can come online in a short time frame, specifically when demand is exceedingly high. There are also relatively abundant supplies of coal, gas, and uranium ore that can be extracted at economical costs. We can extract them, stockpile them, and count on them to provide energy services to the economy's many end users with a great deal of reliability and control. The economy runs smoothly with the assurance of a steady and abundant supply.

On the other hand, we must convert fossil and nuclear fuels via combustion and fission respectively, processes that produce a substantial amount of pollution and/or radioactive waste. They are also finite in supply and not evenly distributed across the globe, which can cause geopolitical problems in times of supply constraint. Improving the efficiency of conversion would necessarily reduce the amount of primary resource necessary to meet an equal amount of electricity demand, which would address supply and pollution issues.

Similarly, using renewable resources that do not deplete, do not produce emissions at point of conversion, and that are often produced locally would also ameliorate supply and pollution problems. The largest current problem with adopting a renewables-only replacement strategy is that renewables are constrained by their respective energy fluxes, meaning the supply is not constantly available. (Using them in a more decentralized method in buildings designed to accommodate them could help with demand pressures and with electricity transmission.)

5.2 OVERVIEW OF THE SECTOR

It is commonly understood that the electric power sector has evolved out of the following three primary elements:

- The supply or source of the electrical energy.
- The delivery system, commonly referred to as "the grid," by which the electrical energy is moved from generators to end users.
- The end users (i.e. customers and rate payers).

The grid itself comprises electrical supply and delivery components: that is, power plants or generators, transmission lines, and distribution through utilities. End user choices or behaviors are what drive supply and delivery decisions. However, the generators and utilities derive profit from electricity supply and delivery sales; thus, there is an embedded incentive for utilities to encourage increased customer demand, at least to a point. An additional widely reported issue with electricity is that, once transformed from the primary energy source, large quantities of it cannot be stored effectively and economically, and thus must be produced at the same time as it is consumed. In traditional, centralized electricity delivery systems, power plants must be producing continuously and operating on demand, meaning that there exists the potential for a large degree of wasted energy.

When considering efficiency, power plants often have well-defined lifetimes of operation, and when they are set to be decommissioned, this provides an opportunity to improve the efficiency of the sector. However, according to Masters, when planning the construction of new power plants, other "demand-side" factors impact the type and size of generator selected. The choice is mostly driven by local and regional electricity demand patterns and by the costs to build, operate, and maintain the plant. Also, the application of some new technologies, especially cogeneration of heat and power, can significantly increase the overall efficiency of a thermal power plant by tapping a portion of the waste heat for space or process heating needs. This may increase the installed costs of the plant, but over the long term, will be balanced by the savings associated with using less primary fuel, especially in areas that have placed strict regulations with fines and/ or levies on emissions, or when fuel prices begin to rise.

5.3 ELECTRICITY DEMAND

Electricity demand represents all the electrical power needed from generation and from delivery to satisfy the various needs and wants that exist in the primary end-use sectors. In terms of measurable units, electricity demand is equivalent to the total watts (W) generated by the power plants and sent to the grid for use by consumers to meet energy needs, consumed over the course of the day, and typically measured in 1000 Wh increments, or kilowatt-hours (kWh).

5.3.1 PEAK DEMAND AND LOAD DURATION CURVES

Although electricity is used throughout the day, it is not used at the same rate all times of the day. The pattern of use changes with activity, typically rising when people wake up and businesses open, and fluctuate in fairly predictable patterns over the course the of the day, associated with human activities. There are well-defined "peaks" in electricity demand throughout the day for a particular region, necessitating enough generating capacity be available to meet these instances. Failure to have enough power to meet peak consumption times would result in power failure. These events, commonly known as "blackouts," happen on occasion and are particularly susceptible to happening when consumers are generating an unexpectedly sizeable power load on the grid, often during a very hot afternoon when cooling systems are running both at residential and commercial buildings simultaneously with other power needs.

To visualize a typical daily electricity demand pattern for the New England region of the United States, let us look at an actual 24-hour demand curve as reported by ISO New England for Tuesday, October 17, 2017, presented in Fig. 5.3. Notice that the least amount of demand occurs around 0300 hours (3 a.m.), presumably when most people are sleeping, and most businesses are closed, followed by a rapidly increasing demand that peaks at around 0730 hours (7:30 a.m.) as people are waking and businesses, schools, and other municipal offices are opening, all "turning on the lights." Demand falls presumably as people leave their homes and stop using electricity in their residences, and the remaining demand shifts primarily to the workplace (and to schools). Another rapidly increasing demand peak happens around 1930 hours (7:30 p.m.) as people return home and turn on the lights and other machines and devices until it is time

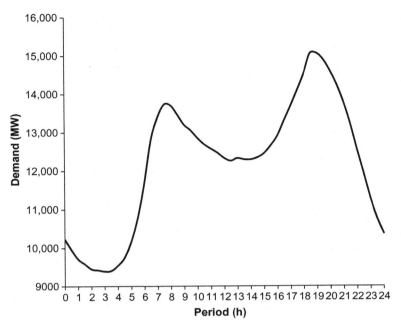

FIGURE 5.3

An electricity demand curve for the New England region of the US during October 17, 2017. This is a power curve. Since it is Megawatts versus hours, the area under the curve is MWh, which is the total energy consumed. This is how total energy and power loads fit together.

Figure created using data presented in ISO New England's online portal, ISO Express: Energy load and demand reports (2017).

to shut down for the night. The second evening peak is higher than the morning peak on this day.

If we now visualize a demand cycle for an entire week, as shown in Fig. 5.4, we can see how demand varies between different days. For the most part, weekday demand is similar from day to day and, as one might expect, weekend demand [periods four (4) through six (6), Saturday and Sunday] is lower than weekday demand, especially the morning peak. And if we do the same over the course of an entire year, plotting every 1-hour period, ordered by time, from mid-October 2016 to mid-October 2017, we see a "noisy" pattern of demand, depicted in Fig. 5.5. Despite the noise, we observe that electricity demand in the summer peaks is higher than in the winter in New England, which makes sense, since air conditioning is used almost exclusively during the summer months in that region (natural gas and oil are the dominant fuels used for heating in the colder months).

Note the largest spikes in demand specifically in the summer months in Fig. 5.5. In these times, a large amount of capacity is needed to satisfy demand at

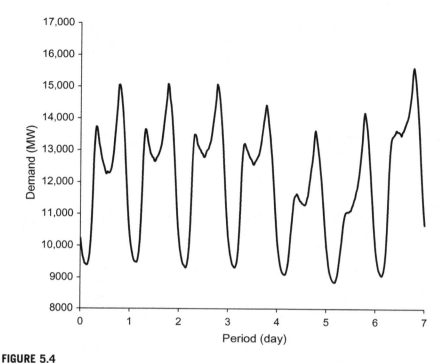

FIGURE 5.4

A demand cycle for the New England region of the US from October 17 to 24, 2017.

Figure created using data presented in ISO New England's online portal, ISO Express: Energy load and
demand reports (2017).

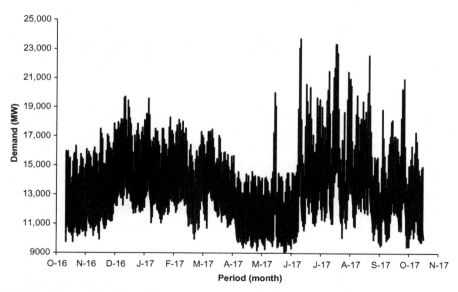

FIGURE 5.5

Yearlong demand cycle for the New England region of the US from October 2016 to
October 2017.

Figure created using data presented in ISO New England's online portal, ISO Express: Energy load and
demand reports (2017).

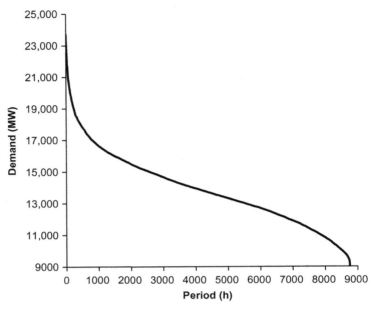

FIGURE 5.6

Load duration curve for the New England region of the US.

Figure created using data presented in ISO New England's online portal, ISO Express: Energy load and demand reports (2017).

these peaks; however, it is only needed for very short periods of time, suggesting a great deal of idle capacity exists most other times. To make better sense of this and the rest of the noisy yearlong demand information presented in Fig. 5.5, it is common to rank the data by the size of the demand versus an hour-long period to produce what is known as a "load duration curve," presented in Fig. 5.6.

Again, the data in Fig. 5.6 is the same presented in Fig. 5.5 but now ranked by demand size. The curve runs out to 8760 hours, the number of hours in a year, however not necessarily ordered chronologically. The highest recorded demand during this 1-year time frame was close to 24,000 MW, while the lowest was around 9000 MW. What is interesting about presenting the data in this fashion is that all peak demand data is on the left and all minimum demand data is on the right, allowing the energy analyst to draw some key conclusions.

For instance, in the first 100 data points of the load duration curve, representing 100 hours of yearly demand, peak power demand drops off by roughly 14%, whereas in the second 100 hours, peak demand drops off by less than 6%. This steepening toward the left-hand side of Fig. 5.6 suggests that there is a significant disconnect between required capacity and the amount of time needed to meet the demand requirement. Analyzing October-2016–October-2017 demand on a percentage basis, we find that 14% of the New England peak capacity requirement was only needed for 1.1% of the year. As we said earlier, this need for excess power for only small periods of time results in a great deal of idle capacity and associated expense. Fortunately, New England's power system is distributed amongst six states, which provides for a better use of excess generation capacity within the grid. Also, forecasting demand and controlling what types of generation need to come on line become important components to a reliable electric power system.

Finally, with cooling being a primary driver of growing demand in all markets worldwide, especially with average ambient temperatures rising in many areas, and with poorer countries investing in electrical infrastructure, it is plausible that more and more markets will come to resemble the New England case. Controlling peak demand events will become more and more important, especially since it is so costly to build new capacity. Incentivizing load reductions by the various customer classes during peak times is one area that can significantly reduce the threat of blackouts. (Building better structures with smaller cooling requirements should also be considered.)

5.3.2 DEMAND VERSUS CONSUMPTION

As we have discussed both worldwide electrical energy consumption and peak power demand in the previous sections, it is important not to confuse the two. Whereas electricity consumption represents the amount of electrical energy that has been consumed over a specific time, in units of Wh (or kWh), electricity demand represents that rate at which electrical energy is consumed for a needed output rating, in units of W (or kW).

Let us consider a classic example to distinguish the two concepts. Say you are operating a lamp with an incandescent light bulb rated at 100 W for 10 hours of time. This would consume $100\,W \times 10\,h = 1000\,Wh$, or 1 kWh of electrical energy. The power plant that produced the electricity to power the bulb needs to generate 100 W of electrical load. Now let us say you are operating 10 lamps with 10 light bulbs each rated at 100 W but now only for 1 hour. This also would consume $10 \times 100\,W \times 1\,h = 1\,kWh$ of electrical energy. However, the power plant now needs to have the capacity to generate 1000 W of electrical load to operate the 10 lamps. That is one order of magnitude more power needed to consume the same amount of energy.

Fig. 5.7 illustrates this point. While total energy required is the same for 10 hours of 100 W demand as for 1 hour of 1000 W demand, the load required from a power plant is very different for these two cases. The reason why it is important to distinguish this difference is that utilities are aware that capacity requirements rise as more and more individual devices get "plugged in" and draw a larger overall load. In response, utilities then charge the consumer for this extra capacity requirement. That is, instead of billing the consumer based on average energy used, they bill based on peak demand, because the utility must be able to

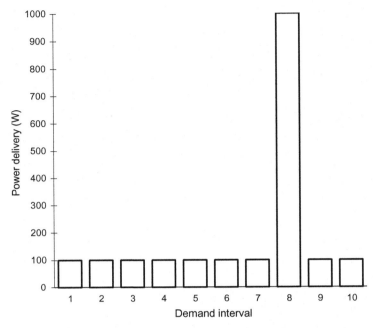

FIGURE 5.7

The effects of peak loads and demand interval durations. In this example, 1000 W of peak load is only required for a short duration compared to most over a given time period. However, a power plant is expected to provide reliable power even at peak intervals, thus necessitating excess capacity to meet this short, extremely large peak event.

reliably deliver electricity to meet the peak need of all its customers, which can strain the total electrical system, especially in time of high total demand.

This is commonly how utilities deal with the commercial sector (and increasingly with the residential sector) and bill them based on utility load cycles of 15 minutes (note that ISO New England provides real-time demand data in exactly 15-minute increments). By billing this way, it encourages customers to reduce demand by turning various loads on at different times or by reducing load requirements. This results in a much more reliable electricity grid.

Alternatively, utilities can bill based on a time-of-use pricing scheme. Going back to Figs. 5.3 and 5.4, peak afternoon weekday demand was between 3 p.m. and 7:30 p.m., which is common for that time of year in New England. Utilities know this and can charge higher prices during these time periods to encourage customers to reduce consumption during those stretches of time. They may also reduce prices during lulls in demand to encourage consumption during times of idle excess capacity. (Note: In a strict sense, Figs. 5.3 and 5.4 are displaying both power demand and electricity consumption. Power would be represented by the curve and consumption would be represented by the area under the curve.)

5.4 ELECTRICITY GENERATOR TYPES

Electricity generating plants vary widely by type and form, but, essentially, they are technologies that input a fuel or other source and output electrical energy to meet end user/customer needs. In addition, as we saw in Figs. 5.3–5.6, customers generally expect electricity to be available all times of the day (there is never a time when zero watts of load are needed), which also impacts generator type.

The baseload power requirement necessitates there be a large portion of generators on the grid that can operate 24 hours/day. Intermittent or "peak" generators provide additional electricity as needed or as available, if coming from a time-dependent energy source. Finally, since electricity is delivered to point of use, these technologies must also be able to adequately integrate with an electrical transmission system or interface directly with a building's electrical systems.

Since electricity must be available as needed, generating facilities must be adaptable and flexible. Different electric generating technologies have very different characteristics in this regard. The purely flow-limited resources have no control in how much is generated with time. Large-scale coal and nuclear power plants have a great deal of thermal inertia and thus are very slow to respond to changes in demand. The electric power generating system requires a portion of the generating capacity to be available on demand. Natural gas turbine power plants can respond much more quickly, but the combined cycle systems that can make them significantly more efficient than coal-fired power plants (CFPPs) inhibit the responsiveness. The electric power sector must be adequately balanced to meet needs both efficiently and effectively.

5.4.1 THERMAL POWER PLANTS

As we have seen in the world power supply of Fig. 5.1 and as was discussed briefly in Chapter 3, Primary Energy Trends, thermal power plants remain the dominant technology used to produce electricity. Most of these power plants operate by burning fuel to convert water into steam that then drives large turbine–generator systems to produce the electricity that gets delivered to customers or rate payers. An overly simplified depiction of this scheme is given in Fig. 5.8. (Water cooling and pollution control systems are not shown.)

Natural gas power plants operate similarly to the scheme in Fig. 5.8, however typically burn the air/gas mixtures directly to drive turbines (operating like jet engines) or use a combined gas turbine plus steam turbine cycle. Nuclear power plants also operate under similar principles but use fissionable material as the source of heat to produce the steam. Thermal power plant efficiency is set by the heating value and quality of the fuel used, and by the individual efficiencies of the boiler, the turbine, and the generator. We review these features within the context of the individual generators in the following sections.

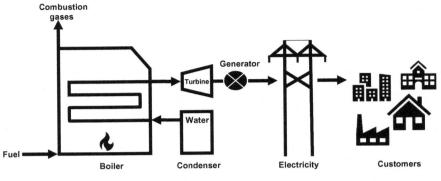

FIGURE 5.8

A simplified depiction of a steam thermal power plant electricity delivery scheme.

5.4.1.1 Coal-fired power plants

CFPPs can employ a wide range of technologies to burn coal for electricity generation, including pulverized coal, cyclone furnace, fluidized bed, and coal gasification/combined cycle combustion. However, according to the World Coal Association, the majority of CFPPs that are currently in operation across the world employ pulverized coal combustion for steam generation, which is the focus of this section. These steam generation plants involve the boiling of large quantities of water and due to its relatively high heat capacity, the generation process is not well-suited for quick starts and stops but rather for maintaining near constant operating temperatures. Thus, they perform more efficiently and effectively as baseload power generators.

The major components of a CFPP are

1. coal storage and preparation,
2. steam generation (furnace and boiler),
3. steam turbine,
4. condenser and recirculator,
5. air emission controls, and
6. electrical generation and power transmission.

(A CFPP may also employ cooling towers to reduce thermal impacts on riverine systems, but this depends on age and location of the CFPP.)

Recall that coal fuels used for power production are classified into four main grades or ranks: anthracite, bituminous, subbituminous, and lignite. As seen in Table 5.1, they vary by how much specific heat, moisture, fixed carbon, ash, sulfur, and chlorine they contain (specific heat is also known as heat content).

In the United States, most of the coal consumed for power generation comes from bituminous and subbituminous sources. Lignite is also commonly used, especially in areas of Europe. Choice of fuel often is dictated by local geologic

Table 5.1 Coal types and average physical properties

Property	Anthracite	Bituminous	Subbituminous	Lignite
Heat content (Btu/lb)	13,000–15,000	11,000–15,000	8500–13,000	4000–8300
Moisture (wt%)	< 15	2–15	10–45	30–60
Fixed carbon (wt%)	85–98	45–85	35–45	25–35
Ash (wt%)	10–20	3–12	≤ 10	10–50
Sulfur (wt%)	0.6–0.8	0.7–4.0	< 2	0.4–1.0
Chlorine (ppm)	340	340	120	120

Table created using source data presented by Indiana Center for Coal Technology Research, Bowen and Irwin (2008).

abundance of the resource. CFPP generator units are typically designed to burn one specific type or grade of coal. Switching to another grade can impact plant performance and overall efficiency. However, choice of fuel can often be related to pollution control/abatement. For example, low sulfur content coal is desirable to reduce particulate emission scrubbing costs, which can motivate a CFPP to test different grades than what the generating unit was originally designed to burn. Also, moisture content can vary widely even within grades, which requires additional energy to dry the fuel before burning in the steam generator.

Coal is typically delivered to the CFPP storage and preparation area by rail. One "unit coal train" is typically 100–130 cars in length, with each car carrying about 120 tonnes (t). CFPP coal consumption depends on the grade of coal being used, the size of the boilers, the efficiency of the plant, and how often the plant operates at the maximum burn rate. The Tennessee Valley Authority reports that small plants can consume as little as 1000 t/day, whereas large plants can consume well over 10,000 t/day. From the coal storage and preparation area, coal is transferred via a conveyor network to fill silos, where the coal is pulverized/milled via crushing and grinding processes. It is then sieved and dried using heated air before being transferred to the furnace and boiler, again via conveyor. (Processing coal to a fine powder increases fuel surface area, allowing for quicker and more complete combustion.)

The steam generator, turbine, and condenser are at the heart of a CFPP and operate following a Rankine cycle, in which water is heated and expanded into steam to drive an external engine. As depicted on the left in Fig. 5.8, the steam generator typically comprises water-filled tubes that line the furnace walls. The originating water source is often a nearby river. The thermal energy released by burning the pulverized coal is transferred and absorbed by the water in these tubes, where it is heated sufficiently to be converted into steam. This steam is then piped into the turbine. The turbine's blades spin, and this mechanical energy is transformed into electromagnetic energy via an attached generator.

Steam returning from the turbine is then cooled in a condenser, and water is pumped back to the steam generator to repeat the process. Additional cold water is brought in from the originating water source, as colder water cools steam more effectively, allowing for more efficient electricity generation. (Cooling towers are employed in newer CFPPs that employ wet recirculation, where once-through cooling systems are no longer allowed, due to excess thermal pollution into ecologically sensitive areas.)

Air emissions are tightly controlled at CFPPs and abatement measures specifically look to control the amount and size of release of sulfur dioxide (SO_2), particulate matter (PM), oxides of nitrogen (NO_x), and mercury. Examples of technologies include dry lime flue gas desulfurization, fabric filter baghouses, selective catalytic reactors, and powder-activated carbon. These systems require energy to run, and sometimes power plants must operate at a lower thermal efficiency to increase pollution control efficiency. Impacts of pollution on plant and animal species as well on human health continually affect the abatement criteria for power plants.

Finally, and as we just mentioned, the mechanical energy of the turbine is transformed into electromagnetic energy via an attached generator, which is then transmitted to the grid. There is very little energy loss between electrical generation and power transmission, and the process of transmission from CFPPs is not particularly different than for other generators. As such it will be discussed later in Section 5.5.

5.4.1.1.1 CFPP efficiency constraints

There are many real constraints on achieving lowest possible heat rate loads at power plants. (Recall in Chapter 4: Energy Flow Analyses and Efficiency Indicators that heat rate is an indicator of power plant efficiency, i.e., the lower the heating rate to generate 1 kW of power, the more efficient the plant.) With respect to the fuel, others have reported that the moisture content impacts sensible and latent heat losses, the ash content impacts heat transfer and auxiliary loads, and the sulfur content affects boiler limits to ensure proper flue gas scrubbing. With respect to cooling, the use of wet recirculation will reduce efficiency versus once-through cooling systems. Ambient conditions (air temperature and humidity) also affect CFPP performance. Emissions control systems also increase on-site power needs and reduce overall efficiency.

Many plant studies have found that efficiency also is significantly affected when power plants operate differently than under their original design conditions. For example, shutting down and starting up the boiler in a baseload plant greatly impacts efficiency. Finally, as a plant ages and moving parts begin to degrade, it will inevitably become less efficient.

5.4.1.2 Natural gas power plants and operation

Natural gas power plants have become increasingly important in electricity generation. Factors that have contributed to this development are increased fuel

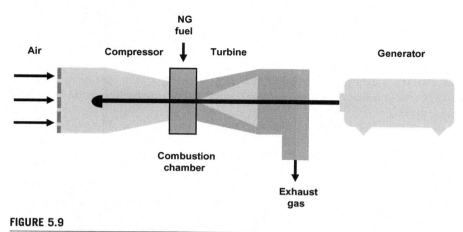

FIGURE 5.9

Simplified depiction of a gas turbine power plant.

abundance (and probably perceived cheaper mid- to long-term prices), relative ease of plant siting and operation, and a much cleaner combustion emission profile. Power plant designs include Rankine cycle steam generation plants similar to coal plants described earlier, simple cycle gas turbine power plants, and combined cycle gas turbine/steam turbine power plants.

According to the US Department of Energy, simple cycle gas turbines have become increasingly utilized for electricity generation, particularly over the last two decades. They are lightweight and reliable, and natural gas is increasingly abundant and more favorable to use for electricity generation due to fast start/stop times for peak power applications and due to a much cleaner pollution profile. In its simplest configuration, such as depicted in Fig. 5.9, a gas turbine plant comprises a compressor and turbine connected to the same shaft that delivers mechanical power to the generator.

In this configuration, compressed air is sent into a combustion chamber where air and natural gas are burned together, converting the chemical energy into thermal energy. The high-pressure combustion gas expands and flows through the blades of the gas turbine, causing it to spin. This mechanical shaft work is then transferred to the turbine generator (since they are connected, when the turbine spins, the generator also spins). The generator converts this mechanical work into electrical energy.

According to Fay and Golomb, the simple gas turbine cycle can best be modeled by the thermodynamic Brayton cycle. In this cycle, the total "net" work produced by the gas turbine power plant is the difference between the work produced by the turbine and the work absorbed by the compressor. The thermodynamic efficiency depends on the ratio of the gas pressures at air inlet and at compressor outlet as well as on the thermodynamic properties of the air and air/fuel combustion products.

Current practical thermal efficiency for gas turbine power plants ranges from 30% up to 37%, with more recent designs surpassing 40%. (As noted in Chapter 4: Energy Flow Analyses and Efficiency Indicators, the current US gas turbine power plant efficiency based on reported heat rates is 30%.) Even at 30%, this efficiency rating is acceptable due to gas turbine's high versatility as peak load generators. Power ratings can range from 50 kW (microturbines) to hundreds of megawatts.

Natural gas power plants become much more efficient when gas turbine generators are mated to Rankine cycle plants that use the hot turbine exhaust to generate steam. These coupled plants are known as combined cycle power plants, where after passing through the gas turbine such as in Fig. 5.9, the hot exhaust gas is sent to a heat recovery steam generator. Here it is used to heat pipes full of water to generate steam, which then flows through a steam turbine, driving a generator (in a fashion like what is depicted in Fig. 5.8 for a generic steam thermal power plant). Combined cycle plant efficiency is on the order of 45%, with newer designs surpassing 50%.

As was alluded to earlier, natural gas has become much more widely used globally because natural gas combustion does not produce a significant (or any) amount of sulfur, mercury, and PM. And the fuel itself is much more uniform in composition than coal: methane being the primary constituent. Burning natural gas does produce NO_x but at much lower levels than coal and even gasoline and diesel used in transportation. To avoid the potential for producing smog locally, natural gas plants require an exhaust stack. Water is also needed at combined cycle facilities. Finally, an increasingly important issue has to do with natural gas production at fields, where the direct leakage of methane into the atmosphere can cause many potential pollution hazards, including methane oxidation, infrared absorption, and warming in the atmosphere.

5.4.1.2.1 Natural gas efficiency constraints

Perhaps the most important parameter that affects thermal efficiency in natural gas power plants, both in simple and combined cycle plants, is temperature. In simple cycle plants, inlet turbine temperature affects efficiency the greatest. According to Fay and Golomb, efficiency can be increased to a small degree by using heat exchangers between the hot turbine exhaust gas and the compressed gas that enters the combustion chamber. However, another limiting factor is the turbine blades themselves, as the materials that comprise them have high temperature tolerances that must not be exceeded. In combined cycle power plants, the largest limiting efficiency factor is the temperature of the exhaust gas, which is lower than a normal steam boiler in a steam-only plant.

Another method to improve fuel efficiency versus thermal efficiency is to configure power plants as heat cogenerators. These are known as combined heat and power (CHP) plants and produce electricity and useful thermal energy from a single fuel source. These stations produce electricity and with a heat recovery unit can also produce steam and/or hot water to be directed to nearby buildings to provide for heating or cooling needs. CHP plants can be highly efficient at using the fuel source with US EPA estimated plant efficiencies approaching 85%.

5.4.1.3 Nuclear power plants

Nuclear power plants are placed into the same class as thermal power plants since they produce steam to power a turbine−generator set to produce electricity. The primary difference is the type of fuel used for the heat source. In the nuclear case, heat is created by the uranium fission reaction. There are several reactor types used around the world, but the two dominant ones are pressurized water reactors (PWRs) or boiling water reactors (BWRs).

Like steam plants in this class, nuclear plants operate on a Rankine cycle. According to the US DOE, for radiation safety reasons, steam pressures and temperatures are kept lower than in fossil plants, with BWRs operating at 30% plant efficiency and PWRs operating closer to 33%. This is an additional constraint to Rankine cycle efficiency constraints described earlier. Supercritical water reactors are a conceptual reactor that could push efficiency up to over 40%.

The overall efficiency of nuclear power plants would need to be evaluated very differently for breeder reactors, since the breeder reactor technology permits tapping the much larger source than conventional nuclear, which includes the 98% of uranium that is U238, rather than being restricted to the 0.8% that is U235. It brings us back to the issue of boundaries discussed earlier. Does the energy input include the energy content of the uranium that is not naturally fissile? If so, then conventional nuclear power reactors would see much lower system efficiencies. If not, and the two kinds of nuclear power are evaluated on the same basis (since they are based on the same uranium feedstock), then breeder reactors would stand to demonstrate efficiencies much greater than 100%.

5.4.1.4 Concentrating solar power plants

Steam generation for electricity production can also be achieved via the utilization of solar energy. However, since sunlight is a diffuse energy source, it cannot achieve high enough temperatures on its own to power heat engines for electricity production but rather must be concentrated. As mentioned in Chapter 3, Primary Energy Trends, this can be accomplished via many technologies, including parabolic troughs, power towers, linear Fresnel systems, and dish/engines, with parabolic troughs, a steam generation technology, representing most of the current production worldwide. All of these technologies achieve adequate concentration by reflecting sunlight against mirrors and focusing it onto a point or line.

According to Masters, in order to achieve utility-scale electricity generation, CSP plants take up a great deal of space. In trough systems, hundreds of collectors are needed and span hundreds of feet each. As such, land requirements are quite high, with trough systems and tower systems requiring 5 acres/MW generated and 8 acres/MW generated, respectively. Also, solar energy varies over the year, so it has been reported that power production can be very different at peak versus annualized average values.

Because of this, Masters explains that efficiency of conversion of CSP systems can be defined on various bases: design peak efficiency, field-measured annual efficiency, or the above described land use efficiency per unit of electrical output. Within the context of the other thermal power plants discussed in this chapter, average annual efficiency may be the most appropriate to consider, and for CSP plants the range is roughly 15%—20%.

5.4.2 WATER TURBINE POWER PLANTS

For established renewable energy sources like hydropower, the need for steam is obviated. Turbine—generator sets are placed where water flows, thus converting mechanical energy directly into electricity. Hydroelectric power is commonly derived from two main sources: the potential energy from high heads of water retained behind dams and by the kinetic energy of flowing rivers (tidal and wave energy are additional lesser utilized sources). The basic idea is to direct water flow through a turbine—generator set to produce electricity, as depicted in Fig. 5.10.

Hydropower is by far the most efficient means of generating electricity on a large scale. What makes hydropower so robust is that energy flows are highly concentrated, and the flow of water through the turbine blades is highly controllable. Also, there are no intermediate energy transformations, with very little heat loss. Because of that, conversion efficiencies can be as high as 95% for large installations and between 80% and 85% for smaller systems.

However, at a renewable power plant, it is common to assume only negligible losses, even if the conversion efficiency was 5%. As many others have pointed out, the reason is that at a hydropower plant, in this instance, one unit of input

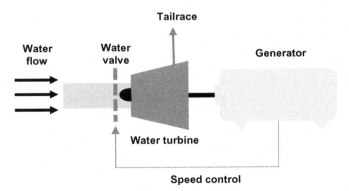

FIGURE 5.10

Simplified depiction of the main features a hydroelectric turbine plant. Water flow would be controlled from the dam reservoir via sluice gates, which is the most common method of hydroelectric generation.

power from water will result in essentially one unit of output electric power. It is not fuel that is being transformed, so measuring how much water was used to produce electricity is less important than measuring the amount of coal, oil, or gas used in the production of power at conventional plants. Of course, it would be advantageous to achieve the maximum possible efficiency since it would reduce the need for multiple sites to produce equivalent power.

5.4.3 WIND TURBINE POWER PLANTS

Wind power is very much like hydropower in that it captures the kinetic energy of a flowing fluid with a turbine—generator set to produce electricity (wind instead of water currents). Essentially, the available power from the wind at any location is dependent primarily on the local wind velocity, then to a lesser extent the size of the wind-blown area (i.e., the total amount of wind hitting a surface), then to an even lesser extent the density of the air. If the object experiencing that passing wind is configured in the right way, it can convert the moving kinetic energy into useful work and ultimately electrical energy.

Horizontal axis wind turbine (HAWT) power plants are the dominant technology for utility-scale production of electricity and their rated output is typically classified by the length of their turbine blades or equivalent swept area such as depicted in Fig. 5.11. Most utility-scale turbines are rated at 2—5 MW and hub

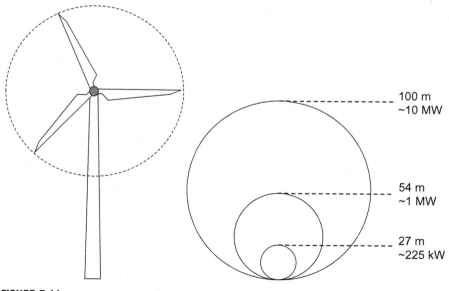

100 m
~10 MW

54 m
~1 MW

27 m
~225 kW

FIGURE 5.11

Depiction of a horizontal access wind turbine alongside a turbine capacity chart (the dashed and solid circles) represents swept areas from the spinning turbine blades. The larger the swept area, the taller the turbine tower (and hub height) must be.

Figure adapted from Ebenhack and Martínez (2013), with permission.

heights of around 50 m above ground, with advances allowing for hub heights approaching 100 m and 10 MW capacity. For context, large coal or nuclear power plants can have generating capacities of 500 MW to 1 GW.

The turbine—generator sets used in HAWTs are similar in design to a hydro-electric turbine plant. The turning blades drive a mechanical shaft connected to a generator that produces electricity directly. The largest difference is that it is virtually impossible to direct the flow of air through the turbine blades, which requires additional safety features to be included in wind turbines, such as a brake between the turbine and the generator to be able to stop the turbine in extremely high winds. (Pitch and yaw control also assist in maximizing wind power energy conversion and in protecting the turbine against high wind damage.)

The efficiency of wind energy to electricity conversion is much less than that for hydropower generation. As mentioned, it is impossible to direct and concentrate air flow in a way that can be done with water flow, thus there are greater losses associated with wind turbines than water turbines. The well-known Betz law suggests that the maximum achievable amount of exploitable wind energy is 59.3%. Electrical generator, subsystems, and availability factor losses also contribute to reducing this number.

In studies related to wind park performance, in which several wind turbines are concentrated in a particular location, it is reported that they can achieve conversion efficiencies as high as 50% at peak, whereas a poorly designed wind park can be as little as 30% efficient at exploiting a wind energy resource. As with hydropower, wind conversion efficiency is not explicitly reported in government data since measuring how much air was used to produce electricity is less pertinent in annual energy accounting. Similarly, it is beneficial to be as efficient as possible to maximize return on initial investment via increased electricity sales using the fewest number of turbines.

5.4.4 GRID-TIED SOLAR PV POWER PLANTS

As we explained in Chapter 3, Primary Energy Trends, photovoltaic (PV) power systems have experienced a tremendous amount of growth in the past decade. In this technology, individual semiconducting cells are linked together into modules that multiply the electricity generated by the sunlight striking them. When these modules are placed into an even larger array, often there is sufficient electricity generated to help power buildings, with the potential to tie into and provide excess power to local electric utility grids.

Grid-tied or grid-connected PV power plants are becoming increasingly important to electricity generation in many countries. These building-scale systems are either pole-mounted or attached directly onto rooftops, and increasingly do not employ battery backup. According to Masters, the PV modules generate direct current (dc) power, so to feed into the building and to provide power to the grid, a "power conditioning unit" or "inverter" must be used to convert the dc power to alternating current power. In these grid-tied systems, where no battery backup is employed, if the PV array is supplying less power than what is being demanded by the building, then the controller will draw from the utility grid to supply the difference. On the other hand, when the array supplies more power than is needed, the excess is sent back onto the grid. The grid, in effect, acts as a battery for each system, taking excess power during off-peak consumption times, when residences are unoccupied, but the array is still producing electricity when sunlight is at its peak insolation.

Based on spectral limitations, the maximum reported efficiency of silicon PV cells is roughly 44% and installed systems typically produce at 10%−20% efficiency. Also, arrays that are designed to track the solar path can harvest and convert much more energy throughout the day than stationary arrays set at one angle. The panels used in these arrays are not any more efficient, but rather the PV systems as a whole are much more effective at producing electricity (i.e., there is greater efficiency of harvesting the resource).

5.4.5 GEOTHERMAL POWER PLANTS

As we described in Chapter 3, Primary Energy Trends, geothermal resources are simply geologic reservoirs from which the "earth heat" can be economically extracted. Geothermal reservoirs can be tapped by drilling a production well into it (similar to drilling for oil and gas) and allowing the hot geothermal fluids to flow through pipes to a power plant. Reservoirs contain heat in the fluids that fill the pore spaces of the source rock as well as the heat contained within the rock itself.

Recall that the fluids that are used for producing electricity are typically found at depths of 1 mi ($\sim$1.6 km) or more, whose energy is utilized by three basic power plant designs: dry steam, flash steam, and binary cycle. Choice of plant design depends on the fluid phase of the reservoir. Dry steam plants use high-pressure steam directly to turn a turbine−generator set. Flash steam plants draw high-pressure hot water up to the surface and run it through low-pressure tanks to flash vaporize the water and then run it through turbine generators (it is possible to employ single, double, or even triple flash configurations). Binary cycle plants, similar to flash steam plants, passes (via heat exchangers) hot water from the reservoir by a secondary fluid with a lower boiling point temperature (typically a short chain hydrocarbon). This secondary fluid vaporizes and is then run through turbines. (Hybrids of the three main designs are also possible.)

Geothermal conversion efficiency is among the lowest of all thermal power plants. This is mainly due to various losses through the reservoir as it makes its way to the power station. According to Zarrouk and Moon, efficiency estimates range from 1% to 20%, with a global average of 12%. Specific factors affecting overall efficiency include reservoir size and fluid temperature, choice of power plant, gas content, dissolved mineral content, as well as ambient conditions. Some of the most pessimistic analyses of geothermal efficiencies make the mistake of counting the thermal losses observed in wellbores in the earliest days and apply them through the life of the field. However, these early losses warm the formations surrounding the wellbore, which decreases the future heat losses. This can be seen by the fact that more mature geothermal fields have continued to produce much longer than some energy experts have predicted.

The boundary analyses for input energy are also challenged by geothermal systems, in that, unlike fossil fuels that exist in finite and well-defined amounts, the heat in reservoirs exists not just in the natural steam, but in the rock materials themselves—and in all of the surrounding rocks as well. As soon as steam production removes some of the heat from the reservoir, the temperature declines slightly, allowing more heat to flow into the reservoir, from all directions. This increases the amount of heat ultimately being tapped. As we have said before, regarding solar PV systems, we suggest that the input energy be counted once it is anthropogenically controlled, as they already routinely are for the fossil fuels. Efficiency calculations are best done on an even basis across the variety of energy systems we employ.

5.5 ELECTRICITY TRANSMISSION AND DISTRIBUTION

Once electricity is generated at the power plant, it must be transmitted and distributed to the end-use demand sectors, which are typically industrial complexes, homes, and businesses (depicted simply on the right-hand side of Fig. 5.8). First, the power plant generates electricity, where it travels to transmission substations. Next, the electricity passes through voltage step-up transformers to prepare it for long distance travel to the demand sectors. As the electricity approaches its destination, it passes through a power substation, then a voltage step-down transformer to deliver lower voltage electricity via distribution lines. It is stepped down again through a transformer drum on "power poles" to deliver safer, low-voltage electricity to the end user.

Transmission is the part of electricity delivery that moves bulk electricity from the power plants to substations closer to areas of the electricity demand. It is commonly known that over these long distances, it is more efficient to transmit electricity at a much higher voltage (and lower current) than what is safe to use in every day applications. This electricity must be reduced or "stepped-down" to lower levels by distribution nodes at the point of end use. Distribution, then, is the last point of delivery to the consumer, where voltage has been adequately reduced to power lighting systems and electrical appliances. In a country as large as the United States, in which electricity systems have been developed since the early 1900s, multiple interconnected grids exist, consisting of hundreds of thousands of miles of high-voltage power lines and millions of miles of low-voltage power lines.

According to the US DOE, efficiency of transmission is affected by ambient temperature, conductive resistance of the transmission wires (which are affected

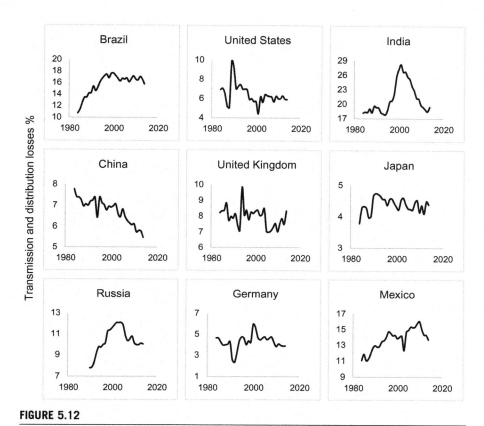

FIGURE 5.12

Energy transmission and distribution losses for selected nations from 1980 to 2014.

Figure created using data presented in World Bank's World development indicators data catalog: Power and communications transmission and distribution losses (2015).

by voltage and current), and electricity demand. That said, transmission and distribution are quite efficient when compared to power plant efficiency. It is common to lose roughly 2% of the electrical energy as heat during transmission, due to resistance in wires. An additional 3%—4% is lost in distribution, as delivering lower voltage electricity is less efficient. There can also exist parasitic losses at transformer drums and substations that result in additional transmission and distribution losses.

In Fig. 5.12, which uses data reported by the World Bank for electricity delivery for a handful of representative countries, we see that most developed nations have transmission and distribution losses that range from 4% to 8%, while in rapidly developing countries, losses can be as high as 29%. As we just mentioned, the variation has much to do with voltage and current delivery differences between nations, distance between generators and end users, and local ambient conditions. The largest losses may also be attributed to poor infrastructure, which may result in parasitic losses, as well as electricity theft, which affects all nations, but disproportionately with the poorest.

As can be seen in Fig. 5.12, in general, as economies develop, grids are likely to extend greater distances from the early central power generation facilities, which tend to incur additional losses. However, further development can include deploying better, more efficient technologies and more locally cited generating capacity can improve the efficiencies.

5.6 ILLUSTRATED CALCULATIONS OF ENERGY FLOW THROUGH VARIOUS ELECTRIC POWER GENERATING SYSTEMS

Electric power generation is unique amongst energy consuming sectors in that its product is a new form of energy. It is the largest single energy conversion, because the product is so versatile and powerful. In the cases of commercial electric power production, it is only a step in the conversion process to deliver useful and effective energy to the consumer. A vast majority of electricity is generated by large, centralized power plants. For this reason, we refer to electric power as primarily a pass-through sector.

Since electric power plays such a significant role in modern, industrialized societies, it is worthwhile to delve into some of the specifics of efficiency losses within this pass-through sector. For selected energy conversion technologies, we illustrate the efficiency losses, based on the initial energy content received at the device or facility. Similar to the macroscopic views of efficiency losses from source to end use illustrated in Chapter 1, Introductory Concepts, we show the losses and the potentially useful energy passing through each step of the process.

For each of the following examples, we take the approximate 15 GW of peak demand for the New England regional power system as seen in Fig. 5.3. Any energy system needs to be designed to be able to meet anticipated peak loads. Therefore, we assess the energy flux through the system by starting with that peak power demand figure of 15 GW and working backwards, through efficiency losses at each step, in order to calculate the necessary energy input for that step. This ultimately leads to an estimate of the energy rate that must be delivered by the electric power facility. The calculations themselves are quite simple: merely multiplying the efficiency for each step together, just as we did in Chapter 1, Introductory Concepts.

The graphical presentation is meant to help visualize the losses at each step through the process, which highlights the largest (and smallest) culprits in efficiency loss. The challenge is estimating the losses at each step. It is very likely for multiple small steps to aggregate into a larger step, but it is imperative to be sure that every opportunity for efficiency loss is included. Note also that these calculations are based on power (rate at which energy is demanded), whereas the similar presentations in Chapter 1, Introductory Concepts were based on energy. The electricity sector is commonly referred to as "electric power" because the facilities are designed to meet the rates of demand. Therefore, it is the rate function, "power" that we use in the following sets of calculations.

We have selected some examples that have conversion steps that can be informative to separate out and compare. Wind and hydropower are essentially one-step conversion processes, hydropower and wind power are simply turbines placed in a flowing fluid stream, although the flow is often diverted and concentrated in large-scale hydropower. This makes it more effective and intensive, allowing economies of scale. This is not truly an energy efficiency issue, though.

Nuclear power could be interesting to consider, due to the limited efficiency of conversion of nuclear energy contained in the fuel. Furthermore, the traditional nuclear power processes can be viewed as very inefficient in that the naturally fissile fraction of uranium is so small. These factors are the basis for experiments with breeder reactors, which would greatly improve the effective use of uranium. Breeders have not yet been able to sustain long-term effective use, so we do not include nuclear power in the following examples.

5.6.1 COAL-FIRED POWER GENERATION

Historically, coal has been the dominant energy source for electric power generation. In some regions, particularly the United States, it is being challenged by natural gas, and as noted in Chapter 3, Primary Energy Trends, it is also being challenged by solar and wind. Coal is likely to remain an important energy source, especially for the electric power sector for many years. However, the process is governed by limits to thermodynamic efficiencies.

Fig. 5.13 illustrates the efficiency losses as power (i.e., the rate of energy flow) moves through a typical CFPP. We first begin with the energy content of the coal received at the plant. Then, after combustion, the energy is sent the boiler, where heat losses occur. Finally, the steam generated at the boiler enters the steam turbine/generator set to generate electricity. It is abundantly clear that the major loss is in the turbine. This is largely because a CFPP relies on a thermodynamic cycle whose efficiency is limited by the ratio of the absolute temperatures at inlet versus outlet. Technologies to improve the combustion efficiency will have relatively small effects on the overall system efficiency. Because of these inefficiencies, in order for such a CFPP to be able to meet a peak demand of 15 GW, a flow of a little more than 35 GW of fuel must be delivered to the plant.

5.6.2 NATURAL GAS TURBINE GENERATORS

In natural gas systems, rather than firing a boiler to generate steam to run a turbine, the hot combustion gasses are fed directly into the turbine. To make this work, the inlet gasses are compressed, so that the exiting combustion products are

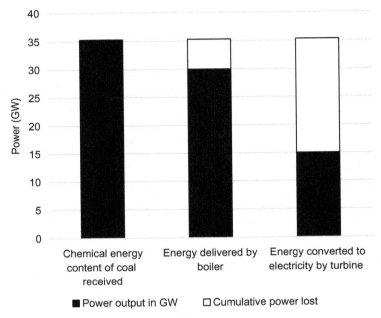

FIGURE 5.13

Power flow through a coal-fired electric power plant to achieve 15 GW peak output. Efficiency losses through the main process steps in a typical coal-fired power plant occur at the boiler and through the turbine.

also at very high pressure. This makes the gas turbine very effective, but the gasses still exit at a very high temperature. Therefore, the efficiency, similar to a CFPP, is still likely to be below 35%. The heat can be tapped in a secondary cycle to generate steam, which can turn a secondary turbine to produce more electricity, thereby increasing the total amount of energy converted to electricity and, thus, increasing the overall efficiency. This process the "combined cycle." It can make natural gas-fired turbines the most efficient conversion technology for a combustion energy source.

Fig. 5.14 illustrates how much less energy input is required to achieve the needed 15 GW of production with the implementation of a secondary cycle. With a single cycle gas turbine, the net efficiency is comparable to (or even potentially lower than) a reasonably modern coal power plant; therefore, it is likely to require more than a 45 GW rate of energy flow in the form of natural gas to be able to achieve 15 GW of peak demand power. While this represents a rather low efficiency, single cycle natural gas turbines are able to respond very quickly to increase or decrease their output in response to changing demand (called "load following.") Thus, single cycle gas turbines are often used as "peaking plants"— only brought online to respond quickly to increased demand.

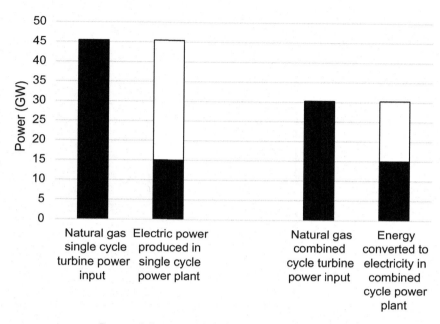

Comparison of power flow through both single and combined cycle gas turbines.

A combined cycle natural gas turbine, though, is much more efficient, since it employs a second cycle, capturing heat from the emissions stream and transferring that to heat a second fluid to the vapor phase, to run through a conventional steam turbine, which generates more electricity from the heat produced by the original fuel. Conversion efficiency increases to around 50% and the power input reduces from 45 to 30 GW. The combined cycle cannot respond as quickly to changes in demand, so cannot be used as a "peaking plant." However, it is possible to operate a combined cycle gas turbine in single cycle mode during ramp up, but it requires substantial redesign and does not appear to be a widely adopted strategy to combine efficiency and response rate.

5.6.3 SOLAR PV POWER GENERATION

Solar power from PV can be deployed in a wide variety of sizes and configurations, and the two most commonly deployed for grid integration are (aforementioned) roof-mounted or pole-mounted arrays at the residential and commercial building level, or very large utility-scale solar farms. Both types feed into the grid, albeit differently. Collection efficiency is a function of both the efficiency of the individual panel cells and the ability of the array to track with the sun's

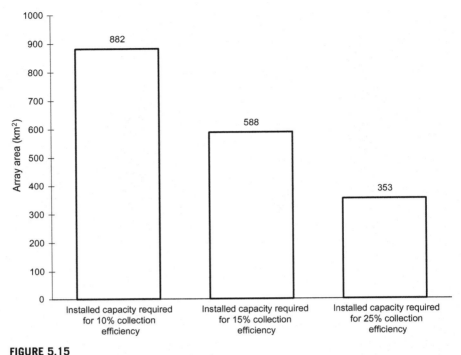

FIGURE 5.15

Required size (in square kilometers) of PV array to meet a peak demand load of 15 GW for different harvesting efficiencies.

path and angle of incidence. These two combined factors can result in total solar collection efficiency (akin to capacity factor) in the range of about 10%−25%. Although we suggest that the amount of solar energy not captured does not represent any inefficiency in the use of anthropogenically controlled energy, it does affect how much area must be covered with PV panels to achieve any given energy production or power output.

For example, if the peak demand of 15 GW were to be generated by a solar PV system, and the system were able to collect the average solar flux of 170 W/ m^2, it would require 88 km^2 of panels, assuming 100% collection efficiency. Now, look at Fig. 5.15, which shows the array area needed at 10%, 15%, and 25% conversion efficiency. All else being equal, the lower efficiency PV system will require a proportionately larger area to produce a given amount of power, which will also only be achievable at peak operating conditions (clear direct sun).

Although there is no expense or human intervention associated with the sunlight not captured and transformed by the PV cells, the incident solar energy is measured per unit area. Therefore, the less efficient PV systems require proportionally more space. We can see that a low efficiency system, converting 10% of the received sunlight to electricity would need more than 880 km^2 of PV arrays installed. For reference, that's about 330 mi^2, or the full area of nearly 10 million homes' south facing roofs of typical 2000 ft^2 homes found in the United States. For a solar PV system with a high collection efficiency, the numbers drop by the ratio of efficiencies. The required space would fall to about 350 km^2. The space requirements, though, do remain very large. Therefore, the efficiency of capturing sunlight does have real significance through its role in establishing the spatial requirements for PV installations. If space becomes a constraint, this may be important.

Of course, if the solar power were stand-alone, it would necessarily be coupled with storage—and the storage system could offer the added value of responding to peak demand loads, obviating the need for a solar PV system with installed capacity based on peak demand. However, the installed capacity must probably be even larger, as (in stand-alone operation) it must be able to produce enough electricity during sunshine to meet energy demands throughout the day and night.

5.7 CONCLUSION

The electric power sector, although not an end-use sector, is perhaps the most important for pass-through energy within modern economies. And we expect it to become increasingly important, in part because wind, solar, geothermal, and hydropower sources are primarily being tapped in modern times to produce electricity. If the transition away from combustion fuels, including fossil fuels, is to proceed, more of our tasks will need to shift toward electricity.

Since this is a pass-through sector, which provides energy to all of the end-use sectors, the efficiencies of electric power production play an important role in the overall efficiency of any system. The largest opportunity for improvements in this sector would come from cogeneration to produce additional work from what would otherwise be rejected as waste heat. A great deal of research is going into improving solar PV production; improvements will no doubt continue to be seen, but it should be remembered that, since the inefficiency is in harvesting inbound sunlight—a free good—the cost of the improvement must compete even more rigorously in terms of price to gain market share. If an inefficient PV cell that costs one-third of another cell that is twice as efficient, the inefficient cell is likely to win in the marketplace.

Perhaps the key challenge for the efficiency of the electric power sector stems from the need to meet peak power demand. This requires systems that can rapidly increase the flow rate of energy produced to meet rising demands (or decrease it to respond to falling demand in off-peak times). Large-scale CFPPs have a great deal of thermal mass and are, thus, incapable of responding quickly to changes in demand. Although natural gas-fired turbines can respond very quickly to changes in demand, the combined cycle mode which is quite efficient, generally sacrifices the load-following ability.

Finally, efficiency is important to the sector, but electric power is also important to overall efficiency considerations, since it can offer better end-use efficiencies than combustion-based energy uses. The transition to heavy reliance on renewable energy sources will almost certainly demand an expansion of electric power into more extensive use in activities including cooking and heating, but most especially in transportation.

REFERENCES AND FURTHER READING

Books and Technical Articles/Reports
Aubrecht, G. J., & Aubrecht, I. I. (2006). *Energy: Physical, environmental, and social impact.* Upper Saddle River, NJ: Pearson Prentice Hall.
Beghin, G., & Phuoc, V. N. (1981). *Power conditioning unit for photovoltaic power systems. Photovoltaic solar energy conference* (pp. 1041–1045). Dordrecht: Springer.
Blok, K., & Nieuwlaar, E. (2016). Introduction to energy analysis. Philadelphia, PA: Taylor & Francis.
Delea, F., & Casazza, J. (2011). *Understanding electric power systems: An overview of the technology, the marketplace, and government regulations.* Hoboken, NJ: John Wiley & Sons.
Ebenhack, B. W., & Martínez, D. M. (2013). *The path to more sustainable energy systems: How do we get there from here?* New York, NY: Momentum Press.
Fay, J. A., & Golomb, D. S. (2012). *Energy and the environment.* Oxford, UK: Oxford University Press.
Hinrichs, R., & Kleinbach, M. (2012). *Energy: Its use and the environment.* Boston, MA: Cengage.
Martínez, D. M., & Ebenhack, B. W. (2015). *Valuing energy for global needs: A systems approach.* New York, NY: Momentum Press.
Masters, G. M. (2013). *Renewable and efficient electric power systems.* Hoboken, NJ: John Wiley & Sons.
McLean-Conner, P. (2009). *Energy efficiency: Principles and practices.* Tulsa, OK: PennWell Books.
Romer, R. H. (1976). *Energy: An introduction to physics.* San Francisco, CA: Freeman.
Shepherd, W., & Shepherd, D. W. (2014). *Energy studies.* London, UK: Imperial College Press.
Zarrouk, S. J., & Moon, H. (2014). Efficiency of geothermal power plants: A worldwide review. *Geothermics, 51,* 142–153.
Energy Information Sources and Technical Reports
Buongiorno, J., & MacDonald, P. E. (2003). *Supercritical-water-cooled reactor (SCWR).* Idaho National Engineering and Environmental Laboratory Report INEEL/EXT-03-01210. Available online at: <https://inldigitallibrary.inl.gov/sites/sti/sti/2699863.pdf>.
Campbell, R. J. (2013). *Increasing the energy efficiency of existing coal-fired power plants.* Congressional Research Service Report CRS-R43343. Available online at: <https://fas.org/sgp/crs/misc/R43343.pdf >

Central Intelligence Agency (CIA). 2017. *World Fact Book*. Accessed from: ⟨https://www.cia.gov/library/publications/the-world-factbook/⟩.

Department of Energy (DOE), Office of Energy Efficiency and Renewable Energy. (2018). *How a geothermal power plant works*. Available online at: ⟨https://energy.gov/eere/geothermal/how-geothermal-power-plant-works-simple⟩.

Department of Energy (DOE), Office of Fossil Energy. (2018). *How gas turbine power plants work*. Available online at: ⟨https://www.energy.gov/fe/how-gas-turbine-power-plants-work⟩.

Energy Information Administration (EIA). (2017). US electric system operating data. Available online at: ⟨https://www.eia.gov/beta/realtime_grid/#/status?end = 20171024T22⟩.

Energy Information Administration (EIA). (2017). *How much electricity is lost in transmission and distribution in the United States?* Available online at: ⟨https://www.eia.gov/tools/faqs/faq.php?id = 105&t = 3⟩.

Energy Information Administration (EIA), Energy Explained. (2017). *Electricity in the United States*. Available online at: ⟨https://www.eia.gov/energyexplained/index.php?page = electricity_in_the_united_states⟩.

Energy Information Administration (EIA). (2017). *Electricity data*. Available online at: ⟨https://www.eia.gov/electricity/data.php⟩.

Energy Information Administration (EIA). (2017). *Today in energy. Most coal plants in the United States were built before 1990*. Scott Jell, primary contributor. Available online at: ⟨https://www.eia.gov/todayinenergy/detail.php?id = 30812⟩.

Energy Information Administration (EIA). (2018). *What is the efficiency of different types of power plants?* Available online at: ⟨https://www.eia.gov/tools/faqs/faq.php?id = 107&t = 3⟩.

Environmental Protection Agency (EPA), Combined Heat and Power Partnership. (2018). *Efficiency metrics for CHP systems: Total system and effective electric efficiencies*. Available online at: ⟨https://www.arb.ca.gov/cc/ccei/presentations/chpefficiencymetrics_epa.pdf⟩.

Environmental Protection Agency (EPA). (2014). *Energy and environment, electricity customers*. Available online at: ⟨https://www.epa.gov/energy/electricity-customers⟩.

International Energy Agency (IEA). (2010). *Power generation from coal: Measuring and reporting efficiency performance and CO_2 emissions*. Available online at: ⟨https://www.iea.org/ciab/papers/power_generation_from_coal.pdf⟩.

International Energy Agency (IEA). (2018). *Key World Energy Statistics 2017*. Available online at: ⟨https://www.iea.org/publications/freepublications/publication/KeyWorld2017.pdf⟩.

ISO New England. (2017). *ISO express. Energy load and demand reports*. Available online at: ⟨https://www.iso-ne.com/markets-operations/iso-express⟩.

Tennessee Valley Authority. (2017). *How a coal plant works*. Available online at: ⟨https://www.tva.com/Energy/Our-Power-System/Coal/How-a-Coal-Plant-Works⟩.

The World Bank. (2015). *World development indicators data catalog power and communications transmission and distribution losses*. Available online at: ⟨http://wdi.worldbank.org/table/5.11⟩.

Other Online Resources

Bowen, B. H., & Irwin, M. W. (2008). *CCTR basic facts #8: Coal characteristics*. Indiana Center for Coal Technology Research, Purdue University. Available online at: <https://www.purdue.edu/discoverypark/energy/assets/pdfs/cctr/outreach/Basics8-CoalCharacteristics-Oct08.pdf>.

Wilson, R. (2015). How much electricity is lost in transmission? *Blog post.* Available online at: ⟨https://carboncounter.wordpress.com/2015/02/15/how-much-electricity-is-lost-in-transmission/⟩.

Grey Cells Energy. (2015). Load duration curves and peak demand. *Blog post.* Available online at: ⟨https://greycellsenergy.com/articles-analysis/load-duration-curves-and-peak-demand/⟩.

New World Resources. (2017). *Thermal coal.* Available online at: ⟨http://www.newworldresources.eu/en/products/thermal-coal⟩.

Enerdata. (2017). *Global energy statistical yearbook.* Available online at: ⟨https://yearbook.enerdata.net/electricity/electricity-domestic-consumption-data.html⟩.

Holcomb Station Power Plant Project. (2017). *New power plant major components.* Available online at: ⟨http://www.holcombstation.com/technology/power-plant-major-components/⟩.

Havens, A. (2017). The difference between electricity demand and electricity consumption. *Lucid Blog Post.* Available online at: ⟨https://lucidconnects.com/library/blog/electricity-demand-vs-consumption⟩.

Hannes, M. (2010). *Unit coal train frequently asked questions.* Available online at: ⟨http://matts-place.com/trains/coal/coaltrain_basics.htm⟩.

Tebbe, P. (2015). Gas turbine power plant. *Engaged in thermodynamics—Textbook supplement.* Available online at: ⟨cset.mnsu.edu/engagethermo/index.html⟩.

Union of Concerned Scientists. (2017). *How it works: Water for power plant cooling.* Available online at: ⟨http://www.ucsusa.org/clean-energy/energy-and-water-use/water-energy-electricity-cooling-power-plant#sources⟩.

World Coal Association. (2017). *Coal and electricity.* Available online at: ⟨https://www.worldcoal.org/coal/uses-coal/coal-electricity⟩.

Patel, S. (2017). Who has the world's most efficient coal power plant fleet? *Power Magazine.* Available online at: ⟨http://www.powermag.com/who-has-the-worlds-most-efficient-coal-power-plant-fleet/⟩.

NaturalGas.org. (2015). *Electrical use.* Available online at: ⟨http://naturalgas.org/overview/uses-electrical/⟩.

Union of Concerned Scientists. (2017). *Environmental impacts of natural gas.* Available online at: ⟨http://www.ucsusa.org/clean-energy/coal-and-other-fossil-fuels/environmental-impacts-of-natural-gas⟩.

American Electric Power. (2014). *Natural gas technology.* Available online at: ⟨https://www.aep.com/about/IssuesAndPositions/Generation/Technologies/NaturalGas.aspx⟩.

Nowling, U. (2015) Understanding coal power plant heat rate and efficiency. *Power Magazine.* Powermag. com. Available online at ⟨http://www.powermag.com/understanding-coal-power-plant-heat-rate-and-efficiency/⟩.

Wärtsilä. (2018). *Combined cycle plant for power generation: Introduction.* Available online at: ⟨https://www.wartsila.com/energy/learning-center/technical-comparisons/combined-cycle-plant-for-power-generation-introduction⟩.

General Electric. (2018). *Combined cycle power plant: How it works.* Available online at: ⟨https://www.ge.com/power/resources/knowledge-base/combined-cycle-power-plant-how-it-works⟩.

Dutton e-Education Institute. (2017). *Baseload energy sustainability.* College of Earth and Mineral Sciences, The Pennsylvania State University. Available online at: ⟨https://www.e-education.psu.edu/eme807/node/667⟩.

Industrial sector energy efficiency

CHAPTER OUTLINE

Of the end-use sectors, the industrial sector contains some of the most energy intensive activities undertaken by society. In this chapter, we:

- review major subsectors, representative industries;
- describe the types of industrial energy processes, intensities, and their efficiencies.

Energy Efficiency. DOI: https://doi.org/10.1016/B978-0-12-812111-5.00006-8

6.1 INDUSTRIAL SECTOR OVERVIEW

The industrial sector is a key component of both regional and global economies. It is where resources are mined and processed, and materials assembled into every day goods and consumables, ranging from toothbrushes to metal containers, to manufactured construction materials, to pharmaceuticals and fuels. Most of these manufactured goods are vital to the function of modern economies, and they are all heavily dependent on energy. Indeed, every manufactured good contains within it a great deal of "embodied" energy. That is, each good requires energy to create it, including the energy to acquire (e.g., mining, harvesting, crude oil production, etc.) and process the raw materials, and transport it to the consumer, thus containing a theoretically measurable amount of primary energy consumed specific to the manufacture and delivery of that product. As such, the industrial sector is commonly classified as an end-use sector because it uses a substantial amount of end-use energy to create finished goods. Nevertheless, we note that some aspects of the industrial sector are more accurately characterized as throughput, because everyday consumers are more likely to use manufactured products than the material from which they are made. Fuels are a good example; they are manufactured and delivered for use in the transportation sector.

According to the US Energy Information Administration (EIA), the industrial sector can be broken down into three primary subsectors: (1) energy-intensive manufacturing, (2) nonenergy-intensive manufacturing, and (3) nonmanufacturing. These subsectors use energy in many ways: as boiler fuel to generate steam or hot water; as fuels to drive engines and motors; and, as process heat for activities as diverse as refining, metals manufacturing, drying, and cooking prepared foods. As we alluded to above, the industrial sector also uses significant energy sources as feedstocks to make other products (instead of merely burning them as fuels). In these manufacturing and nonmanufacturing subsectors, the predominant energy sources are natural gas, coal, and electricity; however, other energy streams include oil, waste heat, pulping liquor, agricultural waste, tree waste, and other wood and agriculture-related residues.

The EIA reports in its International Energy Outlook that from 2010 to 2017, the amount of total global industrial *delivered energy*[1] rose from 210 quads ($\sim$221 EJ) to just around 240 quads ($\sim$253 EJ), and energy demand in the sector is projected to increase by a rate of 1.2% per year out to at least 2040. Fig. 6.1 depicts EIA's reference scenario projection of global energy consumption in 2020, which includes energy transformed in the electric power sector, as well as total industrial, transport, residential, and commercial sector delivered energy. Of the end-use sectors, the industrial sector is the most energy intensive,

---

[1]Delivered energy is defined by EIA in its International Energy Outlook as the heat content of onsite energy use. This includes the equivalent heat content of onsite electricity use (i.e. not including conversion losses that originate in the electricity sector). It also includes fuel used for onsite combined heat and power/cogeneration.

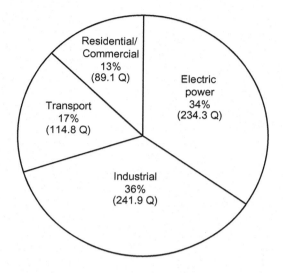

FIGURE 6.1

Projected sector energy consumption in the year 2020. Energy units in quadrillion Btu (Q).

Figure created using data presented in the US EIA International Energy Outlook (2017).

and energy use is on par with the amount of primary energy transformed by the electric power sector.

Also note that of the 241.9 quads (also Q) projected to be delivered to the industrial sector in 2020, only 74.6 quads will come from members of the OECD (Organisation for Economic Co-operation and Development), developed nations that are already postindustrial. Indeed, the majority of industrial activities have been and will continue to be performed in non-OECD nations, primarily China (projected to be around 76 quads) and India (projected to be 15.8 quads). However, as we discussed in Chapter 3, Primary Energy Trends, China appears to have begun a postindustrial drawdown, as evidenced by a large reduction in coal consumption particularly from reduced steel manufacturing in recent years. If this trend continues, this will make India and parts of Africa the most likely regions of continued industrial sector expansion.

Despite short-term drawdowns from China, industrial energy demand as a whole likely will remain high and relatively constant due to an ever increasing and modernizing global population. Because of this continued prominent global industrial energy use, both the US EIA and the OECD International Energy Agency (IEA) have recently begun to better track energy use and intensity within the industrial sector. The express purpose of this increased effort is for modeling specific energy consumption (energy use per unit of product produced) and cataloging appropriate technologies and materials that can be substituted within industrial processes to reduce consumption growth and improve efficiency. As Kesicki and Yanagisawa point out, the biggest impediment to modeling usage trends, and to prescribing changes, is that the sector is not uniform. There are a handful of energy-intensive manufacturing subsectors that consume the majority of the total energy consumed, themselves in very different ways (including using fuels as feedstock), while the remaining subsectors consume the rest of the energy in other equally varied ways. The IEA further points out that there is no one key subsector to focus on for energy reductions such as there is in the transportation sector, for example, where road transport represents most of the energy consumed (nearly 90%) and is a more easily defined target to focus on for efficiency improvements.

Other issues affect how to reduce energy consumption in the industrial sector. For example, the IEA notes with respect to decreasing overall energy demand (or even decreasing growth in demand) within the industrial sector that the most effective strategy for reducing demand is to reclaim waste streams. Recycling previously manufactured goods or using process waste streams can greatly reduce total energy needs by obviating the need to mine, quarry, and/or process virgin material. The issue is that product quality in some cases can be negatively impacted by the use of recycled materials. They note that key physical properties, such as strength, brittleness, brightness, and durability, can sometimes (though not always) be affected by the use of recycled materials. While this may be perfectly fine in most instances, such as in the manufacture of metal cans that hold consumable liquids, when high grade material is required, the opportunity for energy savings is reduced because reclaimed materials would not be feasible to use in those instances. Two additional issues make it challenging to track and subsequently reduce the energy intensity of many industry subsectors. First, waste energy streams, such as the aforementioned pulping liquor, wood waste, and waste heat, are often utilized within combined plant operations. This is not typically tracked adequately and affects how to measure/estimate actual energy demand. Second, industries typically install equipment to be used with certain fuels in mind to produce materials of a specific and/or high quality.

Despite these energy tracking issues, the most likely scenario, as modeled by both EIA and IEA, is that *growth* in industrial energy end use can be substantially controlled by choosing modern and more energy-efficient energy equipment as parts and facilities age past their expected, useful component lives. In the case of rapidly developing and industrializing countries such as India, both EIA and IEA suggest that building new facilities with modern components that have the highest rated energy efficiencies will be the best scenario so that new capacity is already as efficient as possible to reduce future energy demand and at least partially stave off energy price volatility.

6.2 ENERGY-INTENSIVE MANUFACTURING SUBSECTORS

According to a 2015 IEA analysis, the energy-intensive manufacturing subsector, which includes coke ovens, blast furnaces, and petrochemical feedstocks, is responsible for around two-thirds of total energy consumption in the industrial sector. The most important energy-intensive manufacturing subsectors include

- petroleum refining, including feedstocks;
- basic chemicals manufacturing, including chemical feedstocks;
- iron and steel manufacturing;
- aluminum manufacturing;
- cement manufacturing;
- pulp and paper manufacturing; and
- food and beverage manufacturing, including tobacco.

A review of the most prominent industries is in the following subsections.

6.2.1 PETROLEUM REFINING

Crude oil and energy are used to supply raw material, heat, and power to petroleum refinery operations to produce chain-specific hydrocarbon molecules, referred to as petrochemicals, and other nonfuel products. The simplest petroleum refineries perform basic distillations, utilizing differences in density and boiling point of the various molecules within crude oil by heating and vaporizing it in a boiler (or set of boilers) and then separating out desired "fractions" that can then be used for different industrial and end-use applications. The EIA reports that this basic distillation process remains widely utilized by small operators; however, large operators, which dominate the subsector, are much more complex, integrating the ability to convert (or "reform") and blend these hydrocarbons into a much larger profile of chemical goods.

According to a comprehensive 2015 Lawrence Berkeley National Laboratory (LBNL) report on improving efficiency in US oil refineries, the most common gases and liquids that arise from separation and reforming processes are

- refinery fuel gas,
- liquid petroleum gas,
- regular gasoline,
- premium gasoline,
- aviation fuels,
- diesels,
- heating oils,
- lubricating oils,
- greases,
- asphalts,
- industrial fuels, and
- refinery oil.

Each product, or fraction, for the most part comes out as their boiling point is reached, with volatile, small chain hydrocarbons leaving (and captured) first, and the heaviest oils coming off last at the highest temperatures. As described in the report, for modern refineries that distill, reform, and blend crude oil, there are multiple additional refinery process steps that consume energy, including desalting; crude distillation; vacuum distillation; hydrotreating (i.e., hydrodesulfurization); catalytic reforming (i.e., converting naphthas into higher octane products for gasoline blends); fluid catalytic cracking (FCC, breaking long chains into specific smaller chains with a catalyst); hydrocracking (similar to FCC in a high temperature, high pressure hydrogen atmosphere); petroleum coking; visbreaking (i.e., viscosity reduction); alkylation and polymerization; hydrogen manufacturing or steam reforming; gas processing; acid gas removal; and bitumen or asphalt production. An extremely simplified refinery scheme is provided in Fig. 6.2.

All of these processes operate between temperatures of 90°C and 500°C. Some of the product, specifically "still gas" or "refiner fuel gas," is used directly in refinery operations: for generating heat from furnaces; for steam using boilers; for electricity using gas turbines; and for generating hydrogen, which is then utilized in the refinery to remove sulfur and to convert fractions into lighter products. However, additional fuel is purchased to provide for the remaining heat, steam, electricity (much electricity is produced onsite), and hydrogen needed in plant operations. In a US manufacturing energy use study conducted in 2012, it was determined that the majority of the energy used in petroleum refining is onsite, and in US refining, nearly 90% of onsite fuel was applied to process heating, mostly for process unit feed preheaters and distillation reboilers. About 65% was used directly and an additional 23% was used to generate steam, which was then used for additional process heat (again to complete distillation, cracking, reforming, and treating steps). A breakdown of the final energy used by these different processes in typical US refineries is shown in Fig. 6.3.

As seen in Fig. 6.3, the most significant energy consuming processes in refineries are distillation (crude and vacuum), followed by hydrotreating, reforming, hydrogen and alkylate production, and hydro- and catalytic cracking. The US Department of Energy (DOE) notes that energy use will vary in refineries over the course of a year, due to changes in the feedstock, desired product mixes, and sulfur content of the products, as well as operational factors and age of equipment.

The fuel use profile in the refinery subsector is also a bit skewed because of the available energy content within waste streams during the refining process. As seen in Fig. 6.4, the majority of the refinery fuel requirements in US refineries are met by using still (or refinery) gas and petroleum coke. Indeed, a large portion of the process heating needs in refining are supplied by these byproduct fuels. Most of the additional fuel requirements are indirectly tied to offsite electricity and steam purchases, which are significant.

Notable areas within refinery operations that would benefit from energy efficiency improvements include utilities, heat exchangers, fired heaters, process optimization, and motors. All of these can be addressed over time by implementing energy management systems, which track and audit energy use throughout operations with the express goal of reducing energy consumption via better practices and technology investments.

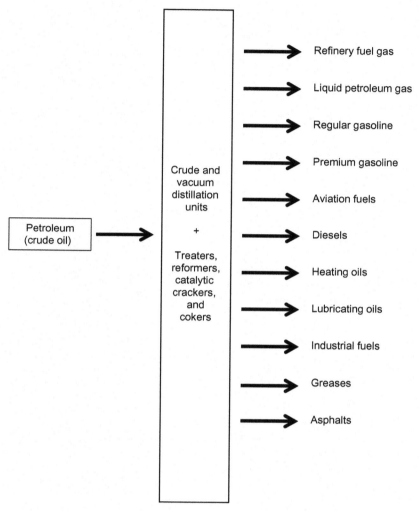

FIGURE 6.2

A simplified petroleum refinery scheme. Products are manufactured from a combination of distillation, reforming, and blending.

Figure created using information presented in Worrell et al. (2015).

6.2.2 BASIC CHEMICALS

According to the Centre for Industry Education Collaboration (CIEC), basic chemicals are produced on the order of millions of tonnes per year. These basic chemicals are divided into petrochemicals, polymers, and basic inorganics and are commonly categorized by their use, by annual production quantities, and

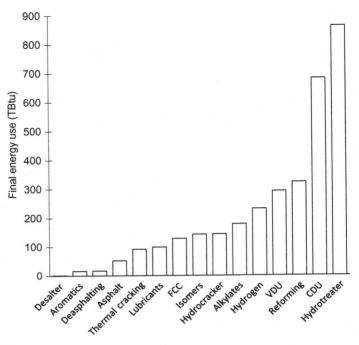

FIGURE 6.3

Final energy use by refinery process in typical US petroleum refineries in units of trillion British thermal units (TBtu).

Figure created using information presented in Worrell et al. (2015).

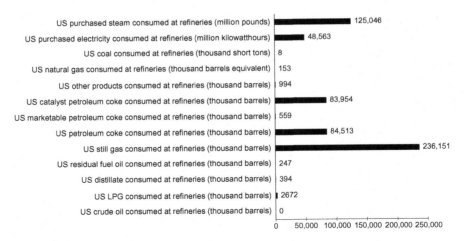

FIGURE 6.4

Fuel consumed at US petroleum refineries in 2017.

Figure created using data presented in US EIA, Fuel Consumed at Refineries (2017).

manufacturing process. The basic chemicals subsector operates much like the petroleum subsector and is another large consumer of hydrocarbon energy. Similarly, a large portion of the energy contained within the raw hydrocarbon material is retained within the final products. These unburned hydrocarbons are feedstocks, used as the "ingredients" in various chemical recipes for end-use consumer products. According to the CIEC, the basic chemicals subsector is the largest subset of the industrial chemicals sector, representing roughly 60% of the total global market.

The straight chain alkane hydrocarbons are found naturally in petroleum and can be separated and converted into more useful products for use within the chemical sector and as end use for consumers. It is common for companies to have integrated chemical plants that produce intermediates and final products. Typical chemical processes in the industrial sector include

- catalysis,
- chemical reactors,
- cracking,
- isomerization,
- reforming, and
- distillation.

Processes are usually integrated to optimize the usage of energy and feedstocks.

Polymers are typically manufactured from petrochemical intermediates, and nonhydrocarbon inorganics are chemicals used in most manufacturing and in agriculture. Common products of all types include plastics, soaps, fertilizers, pharmaceuticals, acids, bases, and pesticides. The 10 top chemical products by annual manufacturing in million tonnes per year as per the CIEC are given in Table 6.1.

Moreover, according to the LBNL, from an energy use/energy-intensity perspective, there are three specific chemicals that require a significant amount of

Table 6.1 Global 10 Top Chemical Products Manufactured Annually by Mass

Chemical Name	Quantity Produced (million tonnes)
Calcium oxide/hydroxide	350
Sulfuric acid	231
Urea	164
Ammonia	146
Ethylene	134
Propylene	94
Benzenes	84
Methanol	70
Sodium hydroxide	70
Nitric acid	55

Table created using information presented in The Essential Chemical Industry — Online (2017).

process energy to manufacture: ethylene, ammonia, and chlorine. According to LBNL, ethylene and its derivative products serve as feedstock for plastics, resins, fibers, and detergents. Ammonia is a key material for the manufacture of fertilizer. Chlorine is used extensively both as a bleaching agent in the pulp and paper industry and as a polyvinyl chloride (PVC) precursor intermediate. Thus these are extremely important manufactured chemicals from a "value-added" perspective.

Ethylene production occurs primarily through thermal cracking and has high energy requirements for the pyrolysis, refrigeration, and rapid cooling steps. Ammonia is produced via an energy-intensive high pressure synthesis of carbon dioxide, hydrogen, and nitrogen. Chlorine production occurs via electrolytic conversion of brine solution into chlorine gas and sodium hydroxide, coproducts of an electricity intensive process. Excluding the energy contained within their respective feedstocks, the specific energy consumption for these three chemicals in the US manufacturing plants based on a 2000 LBNL analysis (Worrell et al.) is estimated to be 26, 16.7, and 47.8 gigajoules per tonne (GJ/t), respectively. A more complete list of LBNL estimated US specific energy consumption for key chemicals (including feedstock energy) is provided in Table 6.2.

Although Table 6.2 represents a 2000 LBNL analysis of chemical industry-specific energy consumption from the 1990s, it served as a good launching point to improve the energy efficiency of the industry of that time. Fuel imports and electricity prices were rising, and the public's environmental concerns were also increasing, causing an industry commitment to reduced energy consumption.

Along similar lines of improving energy efficiency in petroleum refining, the American Chemical Society's (ACS) Presidential Roundtable on Sustainable

Table 6.2 Estimated Specific Energy Consumption for Key US Chemicals

Chemical Product (Including Feedstock)	Specific Energy Consumption (GJ/t)
Ethylene and coproducts	67.5
Ammonia	39.8
Methanol	38.4
Chlorine	19.2
Polyvinyl chloride	11.6
Polypropylene	10.5
Polyethylene	9.3
Polystyrene	9.3
Urea	2.8
Nitrogen	1.8
Oxygen	1.8

Table created using data presented in Worrell et al. (2000).

Manufacturing established a 2020 Vision of the chemical and allied industries with improvements in

1. low temperature heat recovery,
2. low energy separations,
3. cogeneration, and
4. heat and energy management.

As stated previously, much of the energy used by the global chemical industry is embedded in the feedstocks, which, according to the EIA, accounts for roughly 60% of the energy consumed in the sector. The nonfeedstock portion, which is actually used for manufacturing, involves these four high temperature processes targeted by the ACS. Steam is the most important working material; thus, there is a practical need to limit any kind of heat recovery to above the boiling point of water to avoid condensation. If low temperature heat could be better utilized (e.g., through the use of thermodynamic cycles and various working fluids to recover and utilize heat for power generation), this would greatly enhance the efficiency of the chemical industry.

6.2.3 THE IRON AND STEEL SUBSECTOR

The iron and steel subsector is critical to the development of infrastructure and goods in modern economies. Steel, which is an alloy primarily of iron, is strong, malleable, and extremely abundant, thus making it an excellent choice in construction, transportation, and manufacturing in general. Steel also is a major material used to produce mechanical machinery, including automobiles, as well as metal products, and physical infrastructure, such as buildings and bridges. According to the World Steel Association, the world consumed nearly 1.6 billion tonnes of steel products in 2017, double the amount of 2000.

Steel is manufactured primarily by combining pure carbon fuel, or coke, with liquid iron ore to make "pig iron" in blast furnaces, which is then processed into primary steel using basic oxygen furnaces (BOF). This virgin iron comes from iron-rich sedimentary rocks that are mined, transported, and ground into powder. In addition, steel is formed from recycled steel scraps and/or direct reduced iron (DRI), which are manufactured directly using electric arc furnaces (EAF). Virgin-sourced BOF steel, recycled steel, and DRI are typically manufactured separately due to the different processes. Other processes include sintering, casting, reheating, and milling. A basic flow diagram of the processes and raw materials involved in converting feedstocks and fuels to steel products is presented in Fig. 6.5.

Steel mills produce basic steel shapes that are then sent to finishing mills, where the shapes are rolled or hammered into final products (e.g., bars, sheets, or other shapes). Foundries are another integral component of the industry, which produce metal castings with desired final shapes using molten cast iron or steel.

From an energy use/energy-intensity perspective, the main steel manufacturing processes mentioned above require a substantial amount of either raw fuel or

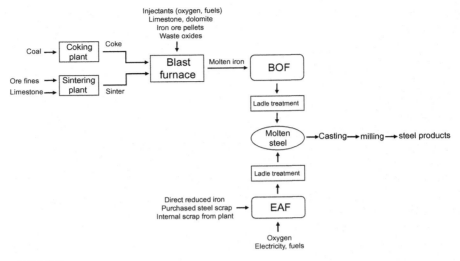

FIGURE 6.5

Simplified schematic of iron and steel production from raw iron materials and fuels to BOF and EAF to casting final products. *BOF*, Basic oxygen furnaces; *EAF*, electric arc furnaces.

Figure creating using a composite of information presented in US DOE (2003) and Worrell et al. (2010).

electricity to reach the temperature needed to melt and form steel products. Estimates of specific energy values, measured in terms of specific energy (GJ/t) or specific electricity (kWh /t) consumption are tabulated in Table 6.3 (based on estimates established by the International Iron and Steel Institute).

As shown in Fig. 6.4 and Table 6.3, most of the energy and electricity end use in steel manufacturing is consumed by furnaces. Indeed, as the 2004 study by the US DOE found in the US iron and steel industry, roughly 81% of energy use was consumed by furnaces, or "fired heaters," with an additional 7% each by boilers and by "motor driven units," such as rollers, pumps, fans, and other equipment. In addition, the 2004 study found that of the onsite energy used in steel manufacturing, roughly one-fourth of it was lost in conversion, distribution, and in motor driven units. Additional offsite losses were due to electricity generation.

Trends in energy intensity have indicated a sharp downward trend between 1950 and the mid-2000s. According to an analysis done by the American Iron and Steel Institute, energy intensity in US steel declined from nearly 60 MBtu/t (~70 GJ/t) in 1950 to about 12 MBtu in the mid-2000s. This is attributed to an increase in the share of EAF production and an increase of the use of more scrap steel in BOFs. Energy efficiency improvements occurred as a result of adopting 100% continuous casting and as a result of new EAF construction, which are inherently more efficient, since it reduces the need for consuming coke and coal. The scrap/EAF production route is inherently more efficient. The use of oil as fuel has also been substituted with natural gas in many new plants, which have

Table 6.3 Estimated Specific Energy Consumption for Key Steel Production Processes

Process	Fuel (GJ/t Product)	Electricity* (kWh/t Product)
Sinter	1.624	28.6
Coke	3.248	39.6
Hot stove	1.74	—
Blast furnace	11.6	—
BOF	0.928	25.3
EAF**	0.464	441.1
Casting	0.0464	8.8
Reheating	1.276	6.6
Hot strip mill	0.0116	133.1

BOF, *Basic oxygen furnaces; EAF, electric arc furnaces.*
Equivalent heat content of electricity is 1 kW h = 3.6 MJ.
**Energy requirements from recycled steel are mostly utilized by the EAF process, whereas virgin steel requires many additional processing steps.*
Table created using source data presented in Worrell et al. (2010).

also increased the efficiency of energy use. Finally, steel manufacturing is greatly impacted by fuel costs, there is a very strong incentive to use the most efficient fuels and technologies, to make a concerted effort to recover waste heat, and to use steel scraps when feasible (BOF uses about 25% scrap, whereas EAF uses almost only scrap).

Finally, the top six steel producing nations in 2017 were China, Japan, India, the United States, Russia, and South Korea at 831.7, 104.7, 101.4, 81.6, 71.3, and 71.0 million tonnes, respectively. China is by far the largest producer, and according to a 2017 report by He and Wang, predominantly uses the BF-BOF production route. As they likely increase scrap/EAF production processes and mimic the US steel industry, it should be expected that global energy intensity will decline. Employing high-efficiency technologies in boiler/steam systems and in process heat systems at time of construction is the key. For plants already built, energy auditing and management systems are important tools to identify and implement energy efficiency, where feasible.

6.2.4 THE ALUMINUM SUBSECTOR

Aluminum is another important metal used widely across the economy. It is light, strong, and resistant to corrosion. As a result, aluminum is used extensively in transportation, packaging, building, construction, and electrical industries. Manufacturers produce aluminum primarily from the processing of the earth-mined ore, bauxite, which is converted alumina (aluminum oxide), and subsequently separated into aluminum and oxygen via electrolytic smelters in EAFs. According to the aluminum manufacturing giant Alcoa, it takes roughly 4 tonnes of bauxite to produce 1 tonne of aluminum. As with steel, primary aluminum production is highly energy intensive. The energy intensity of this process is greatly reduced by mixing new aluminum with plant scrap or recycled aluminum, which accounts for a growing portion of new production. (The International Aluminum Institute reports that in 2015, around 27 million tonnes of recycled aluminum were produced from old and traded new scrap, compared with 58 million tonnes of primary aluminum.)

According to a 2007 LBNL report (Worrell et al.) on best practices in energy intensity in industrial subsectors, key energy-intensive elements of primary aluminum manufacturing include bauxite extraction, aluminum production, anode manufacturing, aluminum smelting, ingot casting, rolling, and extrusion. In contrast, the key elements of secondary manufacturing involve only melting and reshaping recycled aluminum, which is much less energy intensive. A simplified schematic overview of aluminum production from primary and secondary sources to finished products is shown in Fig. 6.6.

From a best practices energy use/energy-intensity perspective, the primary aluminum manufacturing processes mentioned above require a substantial amount of raw fuel and electricity. Estimates of specific primary energy values, measured in terms of specific energy (GJ/t) or specific electricity (kWh/t) consumption, are tabulated in Table 6.4. These are based on estimates established by the LBNL, assuming the electricity is produced using conventional thermal processes with an operational efficiency of 33%.

As presented in Table 6.4, electricity used in smelting is the most energy-intensive component in primary aluminum production. Because the LBNL analysis assumes electricity utilized from conventional power generation, it is possible to

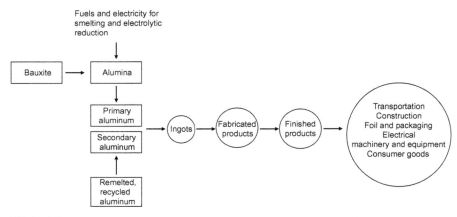

FIGURE 6.6

Simplified schematic of aluminum production from primary and secondary materials and fuels to final products to end-use categories.

Figure created using information and imagery presented in US DOE's US Energy Requirements for Aluminum Manufacturing (2007), with modification.

Table 6.4 Estimated Primary Specific Energy Consumption for Key Aluminum Production Processes

		Primary (GJ/t)	Secondary (GJ/t)
Alumina production	Digesting (fuel)	12.1	
	Calcining kiln (fuel)	6.5	
	Electricity	4.3	
Anode manufacturing	Fuel	1	
	Electricity	0.64	
Aluminum smelting	Electricity	148.4	
Ingot casting	Electricity	1.06	
Total		174	7.6

Table created using data presented in Worrell et al. (2007).

reduce energy consumption by improving the efficiency of the smelting process. According to the LBNL, the Hall–Heroult carbon anode aluminum electrolysis process is the basis for contemporary industrial aluminum smelting and the two chief electrolytic smelting technologies use cells with prebaked anodes and cells with baked in situ anodes, also known as Søderberg cells. Prebaked anodes are far less energy intensive and the subject of continued research and development. Based on thermodynamic limits, the theoretically achievable minimum energy requirement is about 6 kWh/kg of aluminum. State-of-the-art plants can consume as little as 13 kWh/kg of aluminum. Søderberg cell plants consume around 20 kWh/kg. In addition, aluminum production energy intensity can be reduced by improving electricity generation efficiency by using natural gas instead of coal, when possible. Also, in countries with abundant hydropower capacity, aluminum manufacturing facilities are often located nearby, which lowers the energy intensity significantly, due to fuel switching.

Finally, using the United States as an example of energy reduction potential in the global aluminum manufacturing subsector, the massive reduction in energy intensity of the United States industry over the past 40 years has been the result of better, "inert" anode technology and increased utilization of scrap and recycled aluminum. Indeed, according to the US EIA, the energy needed to process and manufacture aluminum from recycled materials is 90% less than from bauxite ore. It will be important to follow how other nations improve energy consumption in the aluminum subsector, especially China, since according to the International Aluminum Institute in 2017, China produced 35.9 million of the 63 million tonnes of aluminum produced worldwide.

6.2.5 THE CEMENT SUBSECTOR

Cement is another important energy-intensive subsector in modern economies. Cement is the key ingredient used in concrete manufacturing, which in turn is used for building highways, streets, and bridges. It is also used for the construction of commercial, residential, and municipal buildings, as well as for public works infrastructure, such as water and waste systems and other utilities. According to Global Cement Magazine, in 2017, 141 countries and territories produced cement in integrated facilities, with an additional 18 countries producing cement from imported clinker. The top five countries of cement production in 2017 were China, India, United States, Russia, and Vietnam. China's estimated cement capacity is between 2.5 and 3.5 Gt/year, roughly equaling the total cement capacity of the rest of the world. Total estimated cement production in 2017 was 4.1 Gt.

According to the Portland Cement Association, cement is comprised mostly of limestone and clay, and basic chemicals including calcium, silicon, aluminum, and iron. "Portland cement" is the general term used to describe most modern cement types, whose name is derived from the original quarries located on the Isle of Portland. The Portland cement process consists of three main steps: (1) raw material preparation, where a primarily limestone and clay mixture are crushed in roller mills to form a fine powder "meal;" (2) clinker production, which consists of sintering the limestone and clay in rotary kilns at combustion gas temperatures exceeding 1800°C; and (3) finish grinding, where the clinker is crushed, ground into powder, and blended with additives to make cement powder. The basic process is depicted in Fig. 6.7.

Cement manufacturing is extremely energy intensive, the most energy-intensive step being clinker production, comprising 80%–90% of total energy consumption and virtually all fuel use, due to the high temperatures required for sintering. Energy needs in this step include preheaters, calciners, kilns, and coolers. Depending on the moisture content of the raw meal feed materials, the kiln process can either be "wet," which involves the evaporation of water from the meal (sometimes up to 36% moisture content), or "dry," which requires almost no evaporation due to the extremely low moisture content (as low as 0.5% moisture content). Multistage preheaters are also employed in dry processes, which further reduce kiln fuel needs.

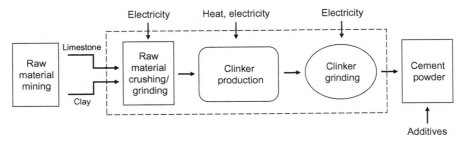

FIGURE 6.7

Simplified schematic of cement production from mining and fuels to final product.

Figure created using information and imagery presented in Afkhami et al. (2015), with modification.

According to a 2008 LBNL report (Worrell and Galitsky), fuel use in a wet kiln process can vary between 5.3 and 7.1 GJ/t clinker, while fuel use in a dry kiln process can vary between 3.2 and 3.5 GJ/t, with a theoretical lower limit of below 2.9 GJ/t. The IEA reports that in 2016, the average thermal energy intensity of global clinker production was 3.4 GJ/t. They also report that although this average number is closer to the lower limit, progress toward lowering thermal energy intensity is still needed in Eurasia, where wet processing persists and average energy intensity remains around 5.7 GJ/t clinker.

According to LBNL and IEA, to reduce energy intensity of cement manufacturing as a whole, the following technologies and practices are recommended:

- increased use of dry-process kilns,
- staged preheaters and precalciners,
- more efficient grinding equipment,
- decreased clinker-to-cement ratios, and
- use of clinker alternatives/additives, including recycled industrial byproducts such as steel slag.

6.2.6 PULP AND PAPER

Paper products are another essential component of a modern economy. Goods often come packaged in paper, tissue, paperboard, and corrugated cardboard. Sanitary tissue is ubiquitous. Even with a move to digital media, paper for printing and writing remains a necessary part of a functioning society. To manufacture these products, industry must convert woody and other cellulosic materials into pulp and roll them into paper. These conversions require a large amount of electrical power and heat (both hot air and steam). An added aspect of the pulp and paper industry is that waste generated during various steps in the pulp and papermaking process produce low-grade fuels that can be collected and used to satisfy internal energy needs, which affords additional opportunities for energy improvement in the sector.

According to the LBNL and the US EPA, the pulp and paper industry comprises pulp mills, paper mills, and paperboard mills that manufacture finished paperboard products from raw and recycled biomass materials. The industry as a whole thus produces commodity grade pulp, primary paper, and paperboard products. From a global production perspective, the Food and Agricultural Organization (FAO) of the United Nations reported world production of paper and paperboard surpassed a massive 406 million tonnes in 2015. Of those, roughly 180 million tonnes were made using wood pulp and 230 million tonnes were made using recovered paper. The top five paper and paperboard producing nations were China (111.1 million tonnes), United States (72.4 million), Japan (26.2 million), Germany (22.6 million), and South Korea (11.6 million).

Major processes that are involved in paper and paperboard manufacturing include raw material preparation, pulping, bleaching, pulp drying, and paper making. Pulp-only mills manufacture what is known as "market pulp," which is produced to be sold on the open market or shipped to another facility where the pulp is used for production of paper goods. Integrated pulp and paper mills produce both pulp and primary paper products at the same site. A simple schematic of the major processes at a wood pulp and paper mill is provided in Fig. 6.8.

Fiber sources also include biomass other than wood, such as straw, bagasse, and bamboo; however, the bulk of pulp and paper processing involves wood as the main fiber source. According to the LBNL, freshly cut timber is typically delivered to mills as logs, where they are cut into smaller sizes, debarked, chipped, and then screened to an optimal size. Sizes smaller than the screen are sent away as "hog fuel," while sizes larger than the screen are rechipped. Once the fiber source is chipped to the proper size, it is then pulped, which is the act of separating the cellulosic fibers from the other parts of the wood. This can be done mechanically, chemically, or by using a combination of mechanical and chemical treatment. Recycled/Recovered paper is also added in this stage. The pulp is then bleached to an industry standard. Chemical recovery is the next step and includes the recovery of pulping liquor, which contains a high enough heating value to use as boiler fuel and for electricity production at the plant. At a pulp-only mill, the pulp is dried for market delivery and paper is made at a different facility, while at an integrated plant, the drying stage is integrated with papermaking. Papermaking steps include drying, pressing/rolling, forming, and stock preparation.

As with most industrial processes, electricity in pulp and paper mills is used to operate motors, pumps, mechanical drives, conveyors, as well as building lighting and ventilation systems. However, the largest use of fuel is in boilers used to

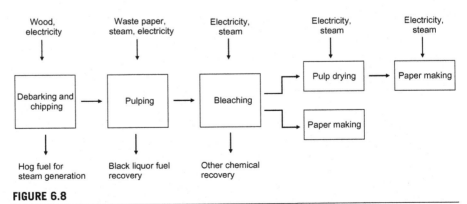

FIGURE 6.8

Simplified schematic of pulp and paper production from raw material and fuels to final product.

Figure created using information and imagery presented in Kramer et al. (2009), with modification.

generate steam for pulping, evaporation, and papermaking. Black liquor, hog fuel, and natural gas are the primary fuels used in the pulp and paper industry. Of course, each mill is different, and energy needs are impacted by the age of the plant, the product mix, and feedstock quality.

Note that because of limited publicly reported information, extensive use of biomass byproducts in pulp and paper manufacturing, and the extensive use of recycled paper and paperboard, it is more difficult to estimate the energy intensity of producing pulp and paper. It is, however, easier to track how much byproduct fuel is produced. For example, the IEA reports that every tonne of produced process pulp produces 19 GJ of black liquor and 0.3–0.7 GJ of hog fuel. Moreover, the IEA suggests a decoupling of energy use and production; however, the IEA, along with LBNL and DOE, suggests that to reduce energy intensity of the pulp and paper industry as a whole, the following technologies and practices are recommended:

- heat recovery to reduce primary fuel conversion,
- cogeneration to utilize waste heat,
- increase recovery and use of biomass waste as fuel,
- increase recovery of waste paper to reduce primary paper production, and
- increased efficiency of steam use and driven systems (pumps, motors, etc.).

6.2.7 FOOD AND BEVERAGE

Food and drink are essential elements of survival, and in the past several thousand years, humans have transitioned from being mostly nomadic foragers to depending entirely on the deliberate planting of crops for food and irrigation for water. Though many humans continue to survive practicing traditional agriculture, most humans now widely utilize mechanization, fossil energy, and agrochemicals to provide for food and drink needs. Humans have created a sophisticated global, and rather energy intensive, industrial agriculture infrastructure providing for billions of people. In fact, a recent report written by the OECD reviewed studies looking at European and United States countries and found that food systems can account for as much as 20% of total energy use in some of these nations. The report goes on to say that much of the growth in energy use is being driven by the increasing demand for processed and "ready-to-eat" foods. The FAO further suggests that globally, end-use energy consumption of food systems is around 30% of the world's total, with more than 70% of the end-use consumption occurring outside of the farm.

The food and beverage sector of United States and EU27 countries is becoming an increasingly important component of national economies, which is responsible for converting livestock, water, and agricultural crops into food and beverage products. According to a recent US Manufacturing Energy Use Analysis report, the food and beverage sector operates using a range of small processing plants producing specific products to large industrial plants that produce hundreds of different products. The major food and beverage subsectors include

- grain and oilseed milling,
- animal slaughter and processing,
- fruit and vegetable processing,
- dairy products,
- beverages, and
- sugar manufacturing.

The food and beverage sector consumes energy mostly in three ways. First, as direct fuel for process heating needed to dehydrate foods, and for combined heat and power and/or cogeneration processes. Second, as electricity for machine drives (e.g., pumps, fans, conveyors, compressors, mixers, etc.), for process cooling and refrigeration, for facilities HVAC (i.e., heating, ventilation, and air conditioning) and lighting, and for other end uses. Third, as steam direct use for process heating, HVAC, machine drives, and other nonprocess uses. Other typical food processing operations include sterilization, pasteurization, washing, cooking, drying, freezing, canning, and packaging, though a wide range of processes are involved in making the thousands of marketed food and beverage products. Unlike other high energy-intensive manufacturing sectors previously described, food and beverage does not generate very much of its own electricity onsite, thus is unable to greatly control efficiency of generation and transmission. The food and beverage subsector does, however, greatly control steam generation onsite. Nonprocess energy includes building/facility HVAC and lighting, and onsite transportation. A simplified plant schematic of a typical food and beverage manufacturing plant is depicted in Fig. 6.9.

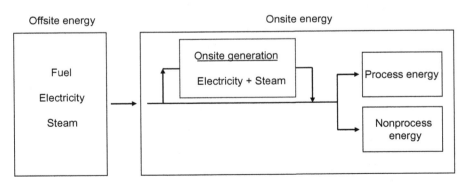

FIGURE 6.9

Simplified schematic of food and beverage process energy flow. The food and beverage subsector process and nonprocess energy flow closely resembles petroleum refining.

Figure created using information and imagery presented in the 2014 US EIA Manufacturing Energy Consumption Survey, with modification.

Table 6.5 Onsite Current Typical Energy Intensities of US Food and Beverage Sector Products

	BTU/lb	MJ/kg
Soybean oil	3688	8.58
Corn oil	1665	3.87
Cane sugar	2526	5.88
Beet sugar	3682	8.57
Fruit juice	741	1.72
Canned fruits and vegetables	942	2.19
Frozen vegetables	1293	3.01
Frozen fruit	776	1.81
Lard, tallow, fat	4573	10.64
Pork products	348	0.81
Red meat products	1141	2.65
Poultry (slaughtered)	91	0.21
Fluid milk	283	0.66
Cheese	1183	2.75
Ice cream	715	1.66
Butter	534	1.24
	BTU/gal	**MJ/L**
Beer	4714	1245.38
Wine	2214	584.95

Table created using data presented in US DOE Bandwidth study on energy use and potential energy savings opportunities in U.S. food and beverage manufacturing (2017).

A recent "bandwidth" energy use study by the US DOE reports that the six bulleted subsectors listed previously account for 80% of the energy used (roughly 1.23 quad or 1.3 EJ) in the sector and over 226 Gt of product. Based on the aforementioned typical energy-intensive processes, onsite energy intensities of various processed foods and beverages are listed in Table 6.5.

Three main technologies for reducing the most energy-intensive processes include energy-efficient drying technologies such as mechanical processing for water removal prior to employment of thermal applications for dehydrating and drying, increased waste heat recovery in food and beverage manufacturing, and increased use of onsite combined heat and power and cogeneration technologies to reduce fuel consumption. Adopting state-of-the-art technologies, DOE estimates that the US food and beverage sector can save 0.35 EJ of energy annually, a reduction of roughly 27%.

6.3 NONENERGY-INTENSIVE MANUFACTURING SUBSECTORS

The commonly reported nonenergy-intensive manufacturing subsectors include

- pharmaceuticals,
- paint and coatings,
- adhesives,
- detergents,
- chemical feedstocks,
- metal-based durables (MBD).

The above-listed chemical subsectors are most accurately described as downstream chemical and chemical product manufacturing subsectors, while MBDs include metal products, machinery, computer/electronic products, transportation equipment, and electrical equipment.

As pointed out by Ramirez, Patel, and Blok in a 2005 article on nonenergy-intensive manufacturing in the Netherlands, there is much less information and emphasis from international agency reporting with respect to energy intensity and efficiency of these important manufacturing subsectors. For example, while not nearly as energy intensive as the upstream chemical industry described in Section 6.2, downstream chemicals such as pharmaceuticals consume and/or represent a significant amount of energy. In the United States, downstream chemical energy consumption is reported to be about 0.8 EJ around 2010, which is slightly smaller than all the energy consumed in the food and beverage sector. Likewise, during the same time period, the MBD subsector consumed roughly 0.8 EJ.

Energy end use in the downstream chemical subsector and the MBD subsector primarily involves electricity and natural gas (including liquid petroleum gas) fuel consumption. These are mostly for direct uses and for driving machines, including

- HVAC systems,
- fume hoods,
- clean rooms,
- motors, pumps, and compressors,
- refrigeration, and
- lighting.

In addition, natural gas is used for process heating, including heat and steam distribution and cogeneration, and in the case of downstream chemicals, for chemical product synthesis.

Energy intensity for the nonenergy-intensive subsector is typically measured on a value-added basis. That is, industries within this subsector are measured by energy use per unit amount of monetary value added to the respective economy, typically reported as Btu or MJ/US Dollar (or Euro, or British Pound, etc.). For example, the EIA reports that in the US MBD subsector, 2010 value added was broken down as follows:

Transportation equipment	737 billion US 2005$
Computer and electronics	406 billion US 2005$
Machinery	333 billion US 2005$
Fabricated metal products	283 billion US 2005$
Electrical equipment, appliances, components	101 billion US 2005$
Total MBD subsector	1.86 trillion US 2005$

MBD industries also reportedly used 942 trillion Btu in 2010, so the subsector's energy intensity was 506 Btu/US 2005$, or roughly 533 MJ/US dollar in 2005 money.

Because the MBD subsector is dependent evenly on electricity and gas, the energy intensity could be improved by better energy efficiency upstream in the electric power sector or by better energy efficiency in plant operations. These fall under aforementioned "energy management activities," which additionally includes energy auditing/assessment, electricity load control, and training, among others. Similarly, the downstream chemicals subsector energy intensity would be assessed by measuring energy use versus value added by how much monetary value they contribute to the economy.

6.4 NONMANUFACTURING SUBSECTORS

One last industrial area commonly reported on is the nonmanufacturing subsector, which includes

- agriculture, forestry, and commercial fishing,
- coal mining, oil and gas extraction, metals and minerals mining, and
- construction.

The commonality amongst these industries is that they form the basis for major economic systems, namely food and materials, fuels and minerals, and physical infrastructure for most of the other sectors of an economy. Enterprises that focus on agriculture, forestry, or mining are often the first step of an industrial supply chain of product to the consumer, or in the case of construction, an intertwined component. Also, companies are less inclined to track and/or report delineated energy consumption if they are also involved in downstream activities.

Following a 2011 report by the PNNL, nonmanufacturing accounts for roughly 14% of US industrial energy consumption, which is a good indicator for energy consumption in an established industrial or postindustrial nation. Moreover, the majority of the energy directly used in agriculture, mining, and construction is for fuels to power the machinery required to cultivate, mine, and build. Total delivered energy used in 2011 was reported to be 2126 trillion Btu and value added was estimated to be 769 billion US 2005$ for an energy intensity of 2764 Btu/US 2005$. This value is much higher than the value determined for the MBD subsector and provides an indication of artifacts that might not fully explain how energy is used in nonmanufacturing. Indeed, the PNNL report explicitly states that it may be incorrect to simply subtract establishment-reported energy consumption in manufacturing from total industrial energy use as published by EIA. The most definitive information that can be gleaned is that fuels use has declined and electricity use has increased with time; however, both mining and construction really are functions of the real-time health of the economy and current demand for energy and minerals.

6.5 TECHNOLOGY IMPLEMENTATION AND ENERGY SAVINGS ACROSS SUBSECTORS

As depicted in Fig. 6.1, the EIA projects that the world's industrial sector will be consuming roughly 36% of all primary energy consumed globally in 2020, roughly consistent with average industrial energy consumption since the year 2000. In addition, using current data reported by the US Central Intelligence Agency, it appears that many developed nations also produce roughly 20% of their GDP via their industrial sectors. As explained previously, the ratio of energy consumption to GDP is a commonly utilized (though not entirely accurate) indicator of a nation's industrial energy efficiency, known as energy intensity, or the amount of energy required to produce one unit of output (e.g., a kilogram of steel or liter of fuel). Thus nations strive to reduce this ratio as much as possible.

Reducing energy intensity is an import concern for national economies, especially when faced with volatile energy prices. As nations seek to remain competitive with other countries, it is the key for their industrial sectors to adopt better energy efficient technologies to reduce energy intensity. As we have seen throughout this chapter, the industrial sector comprises many different types of manufacturing subsectors and many require subsector-specific technology to produce their specific manufactured good or goods. However, as we piece together all of the manufacturing subsectors, there appear to be many established and prospective "cross-cutting" technologies that can be applied across all subsectors to reduce total sector energy consumption.

A joint study between the LBNL and the American Council for an Energy Efficient Economy (ACEEE) in the late 1990s isolated emerging energy efficiency technologies that were likely to be developed and deployed in the first two decades of this century both in the US economy and globally. In the report, they isolated over 50 technologies that had the greatest prospects for energy savings potential and likely adoption. Of those, the LBNL and ACEEE reported a list of technologies with the highest primary energy savings potential, reproduced in Table 6.6. This study remains relevant because businesses tend to operate already installed equipment until its useful lifetime has expired, so the study results indicate likely movement in the industry in the 2020s.

As presented in Table 6.6, it is evident that most of the savings potential in the industrial sector crosses all subsectors, with the majority of savings occurring if subsectors invested in better optimized motor-driven systems, more efficient pumps, and advanced reciprocating engines. Indeed the Kesicki and Yanagisawa report (mentioned at the beginning of this chapter) notes that current industry energy savings are occurring from more efficient technologies, replacing old facilities with modern (and intrinsically more efficient) facilities, and process changes and optimization such as those listed in Table 6.6.

Table 6.6 Primary Energy Savings Potential of Various Existing and Prospective Industrial Energy Technologies

Technology	Subsector	Savings (Tbtu)	Savings (EJ)
Motor system optimization	Cross-cutting	1502	1.585
Pump efficiency improvement	Cross-cutting	1004	1.059
Advanced reciprocating engines	Cross-cutting	777	0.820
Compressed air system management	Cross-cutting	563	0.594
Advanced lighting technologies	Cross-cutting	494	0.521
Advanced CHP turbine systems	Cross-cutting	484	0.511
Advanced lighting design	Cross-cutting	231	0.244
Fuel cells	Cross-cutting	185	0.195
Near net shape casting/strip casting	Iron and steel	138	0.146
Sensors and controls	Cross-cutting	137	0.145
Fouling minimization	Petroleum refining	123	0.130
Membrane technology wastewater	Cross-cutting	118	0.124
Microturbines	Cross-cutting	67	0.071
Electron beam sterilization	Food processing	64	0.068
Black liquor gasification	Pulp and paper	64	0.068
Efficient cell retrofit designs	Aluminum	46	0.049
Process integration	Cross-cutting	38	0.040
Autothermal reforming	Chemicals	37	0.039
High consistency forming	Pulp and paper	37	0.039
Condebelt drying	Pulp and paper	34	0.036

Table created using source data presented in LBNL and ACEEE (2002), with modification.

Finally, we note that there often exists substantial variation within subsectors both within a singular country and across country borders. This is often a function of established local infrastructures and practices, as well as available energy sources and raw materials used by a particular local manufacturer. Moreover, many of these subsectors produce their own heat and electricity and burn their own waste products as fuel. This causes different qualities of final products, as well as different energy consumption needs, thus making it difficult to accurately measure energy intensity of many industries. Despite these nonuniformities across the industry subsectors, it is clear that energy savings can be achieved widely.

6.6 CALCULATION: ENERGY AUDITING AND ESTIMATING PAYBACK FOR ENERGY EFFICIENT TECHNOLOGY ADOPTION DECISIONS IN THE INDUSTRIAL SECTOR

Industrial manufacturers considering investments in energy-efficient technologies for their plants and operations tend to have multiple projects that could be pursued to reduce overall energy consumption and annual expenditures. According to Anderson and Newell, manufacturers tend to adopt energy saving technology more readily for projects with shorter payback periods (the amount of time it takes to pay back initial capital investments in technology through reduced annual energy costs), lower upfront or initial costs, and higher annual savings. They also tend to pursue more projects during times of higher energy/fuel prices and awareness for greater energy conservation. Much of the time, however, manufacturers are busy manufacturing goods; making the best decisions for energy efficiency adoption in the face of upfront costs tends to fall by the wayside. This tendency to put off improvements until a later time (or never) can be mitigated by ensuring manufacturers are well informed of the cost savings potential of new/more efficient technologies to produce the same quality of product. Thus "information programs" are essential elements of improving industrial energy efficiency.

6.6.1 ENERGY EFFICIENCY INFORMATION PROGRAMS AND THE ENERGY AUDIT

Energy efficiency information programs refer to local or national proenergy efficiency policies designed to increase awareness of energy saving opportunities as well as to provide technical assistance with technology adoption projects. These programs include

- professional workshops and training,
- advertising,
- product labeling, and
- energy audits of manufacturing plants.

Of the programs listed, energy auditing (usually government sponsored) has become increasingly important in the adoption of energy saving technologies at industrial plants, often because it provides guided technical assistance, detailed energy use analysis, and cost estimations for technology adoption and for likely savings, which can sometimes be immediate.

According to Kluczek and Olszewski, the energy audit is used to analyze energy flow and assess energy cost savings based on a baseline measurement of current consumption and calculated savings due to the adoption of more efficient technologies and energy use strategies. Indeed, the industrial plant energy audit is an energy planning tool to educate manufacturers on how to implement cost-saving measures through energy efficiency. The structure of an energy audit is as follows: (1) preaudit questionnaire, (2) initial analysis of process, (3) further auditing directions, (4) measurements and data collection, (5) energy analysis and final report, and (6) implementation report.

As per Kluczek and Olszewski, the preaudit questionnaire is probably the simplest but the most important aspect of an audit. It is usually a one-page information gathering document seeking information on the specific manufacturing process, including processed materials, production scale, and final products. These provide auditors with an understanding of the primary energy conversion processes at the plant. In addition, preaudit questionnaires request annual energy usage and costs based on monthly utility bills, including electricity, fuels (oil, gas, coal, etc.), water, and processed gases. This information provides a generic view of the plant's energy conversions and flows, which allows auditors to identify areas for potential savings.

The additional steps in the audit involve meetings and interviews with plant personnel at the plant to better understand the manufacturing process and energy conversions at the facility. This walkthrough allows for further auditing, whereby auditors will decide on specific data to collect, including measurements, document gathering, and inventorying of key manufacturing equipment. Once measurements are made and all relevant data are collected, including additional communications with plant personnel, auditors attempt a more robust energy analysis and generate a report with recommendations to the manufacturer. Finally, the implementation report involves auditors following up 6 months or so after to verify the observations and to see how many recommendations have been implemented and how much energy savings have been realized. A simplified depiction of the audit structure on a generic manufacturing process is presented in Fig. 6.10. The most variable component of the auditing process is most likely the energy analysis, because manufacturing subsectors can vary widely in operation and final product. The analysis is very much a function of the auditor's knowledge and experience across subsectors.

The primary outcomes of an industrial energy audit are recommendations that result in energy savings (or reduced energy intensity) at the manufacturing plant and, subsequently, cost savings associated with the energy savings. Per Kluczek and Olszewski, energy savings are commonly formulated in absolute or relative

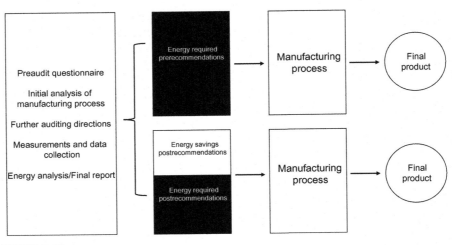

FIGURE 6.10

Energy audit structure on a generic manufacturing process. By implementing different, more efficient technologies, it is possible to manufacture the same final product from less input energy, thereby reducing the energy intensity of manufacture.

Figure created using information presented in Kluczek, A., & Olszewski, P., (2017).

terms as a function of the difference in energy used in production. That is, energy savings is a function of energy actually used in production preaudit and the energy used postaudit, with recommendations implemented.

Cost savings are tied directly to absolute energy savings and can be calculated simply by multiplying absolute energy savings by the energy unit price

As highlighted by Kluczek et al., cost savings play a prominent role for manufacturers when deciding to implement the audit recommendations. Moreover, potential cost savings provided in the audit report allow for determination of payback period—again, the time needed to recoup investment costs associated with project implementation.

6.6.2 ENERGY EFFICIENCY PAYBACK PERIOD

In manufacturing facilities, the different plant processes require different types of equipment that each has a rated and/or measured energy consumption value. As we have described earlier, investing in new, more efficient equipment involves an upfront capital expenditure that will take time to recoup in reduced energy costs. Manufacturers want to know how quickly the return on investment will take to "break even" when making their decision. However, as many experts note, by focusing only on the initial cost, and not long-term savings because of an efficiency improvement, it is likely that the manufacturers will place themselves at financial risk when energy prices inevitably rise in the future. Other factors such as product life and maintenance costs also affect this result. Thus it is important to look at what goes into payback calculations.

Payback period is typically calculated using what is known as "simple" payback, which is

$$\frac{\text{Upfront cost in \$}}{\text{Annual energy savings in \$ per year}} = \text{Payback period.} \qquad (6.1)$$

For example, if a new hypothetical lighting system or relamping program (i.e., swapping out incandescent lighting with high-efficiency LED lights) costs a manufacturer $7000 and the projected annual energy savings are $5000 at current electricity prices, after just 1.4 years ($7000/$5000), the initial cost of purchase has been repaid through the energy savings[2]. It is also interesting to note that many facilities that have already adopted efficient lighting technology do not utilize lighting management features embedded in the systems (i.e., using software to turn lights on and off on a predetermined schedule). Just by activating the software management systems, many facilities can realize immediate savings with no additional investment (but were only made aware through the energy audit process).

Beyond simple payback, it is also important to look at the useful lifetime of equipment. Hanna, Harmon, and Peterson note that when determining the useful life of equipment, one must consider how long the equipment will remain functional (with reasonable maintenance and repair costs) compared to the availability of more efficient replacements. Take, for example, four generic energy efficiency projects recommended to a manufacturer, tabulated in Table 6.7. Projects A and B have the same initial cost and annual savings, resulting in the same simple payback ($10,000/$2500 = 4 years in this theoretical calculation). However, the equipment lifetime in Project B is 10 years versus 5 years in Project A. Thus over a 10-year period, Project B is a much more advantageous choice. Even if Project B was $5000 more costly than Project A, noted as Project C in Table 6.7, the additional 2 years in simple payback is outweighed by the 10-year savings. Alternatively, if a manufacturer decided to choose Project D, which has a much lower initial cost and only a 2-year simple payback, the annual savings from Projects A, B, and C make them better choices for the plant in the longer term.

Based on the example presented in Table 6.7, it is important when looking at recommendations from an energy audit to know specific details of the efficiency improvements. A thorough report should consider these additional variables when making such recommendations.

---

[2]If the new lighting system were more expensive to maintain (it should not be), then that added annual operation cost would be subtracted from the annual energy cost savings.

Table 6.7 Comparisons of Projects With Different Annual Savings and Useful Equipment Lifetimes

Payback Parameters	Project A	Project B	Project C	Project D
Initial cost	$10,000	$10,000	$15,000	$1000
Annual savings	$2500.00	$2500.00	$2500.00	$500.00
Simple payback	4 years	4 years	6 years	2 years
Useful equipment lifetime	5 years	10 years	10 years	10 years
Cost (−) or savings (+) at the end of year				
1	−$7500.00	−$7500.00	−$12,500.00	−$500.00
2	−$5000.00	−$5000.00	−$10,000.00	$0.00
3	−$2500.00	−$2500.00	−$7500.00	$500.00
4	$0.00	$0.00	−$5000.00	$1000.00
5	$2500.00	$2500.00	−$2500.00	$1500.00
6	−$5000.00	$5000.00	$0.00	$2000.00
7	−$2500.00	$7500.00	$2500.00	$2500.00
8	$0.00	$10,000.00	$5000.00	$3000.00
9	$2500.00	$12,500.00	$7500.00	$3500.00
10	$5000.00	$15,000.00	$10,000.00	$4000.00

Model based on Hannah, M., Harmon, J., & Petersen, D. (2011).

6.6.3 CASE STUDY: TEXAS TILE MANUFACTURING LLC

In response to the oil embargo and rising energy costs of the late 1970s, the US federal government established Industrial Assessment Centers (IAC) to help small and medium-sized manufacturers reduce costs from inefficient energy use. Since its creation, IAC has provided manufacturers across the United States with no-cost energy auditing and technical assistance. To date, the program reports that it has conducted over 18,500 assessments with more than 140,000 associated recommendations. These assessments cover all manufacturing subsectors and are available through the IAC online database.

For an example of a typical assessment in the IAC database, consider the vinyl flooring manufacturer, Texas Tile Manufacturing LLC. An assessment was conducted in 2006 at their Houston, Texas facility. Texas Tile produces vinyl-composition floor tile for commercial applications. The raw materials used in the process include polyvinyl chloride resins, as well as plasticizers, pigments, and stabilizers. These materials are mixed, blended, and formed into large sheets. They are then milled to their desired thickness, then cut and boxed for shipment. The most energy-intensive process involves heating the uncut vinyl sheet with natural gas heaters so that the vinyl maintains its pliability as it travels through the production process.

A university-based IAC assessment team performed a routine energy audit at Texas Tile, including a site visit and data gathering with plant personnel. From that audit and visit, they provided the manufacturer with a report containing nine recommendations to reduce energy use and provide cost savings, which is reproduced in Table 6.8. Notice that most recommendations estimated substantial energy and annual cost savings, with immediate or nearly immediate payback. The top recommendation, replacing natural gas pilot lights with spark igniters, was by far the largest energy saver, which had an energy and a nonenergy cost savings effect. As mentioned above, since the vinyl sheet needs to remain pliable throughout the production process, it is heated with gas-fired heaters at various locations throughout the plant, ignited by the pilot lights. Since the lights needed to remain on even when there was no vinyl passing through the system, a significant amount of waste heat was being produced. By switching to spark igniters, which do not require a constant burning of gas, but rather initiate combustion as needed by the production process, this eliminated the need for pilot lights, resulting in an annual savings of 12,100 MMBtu or $113,000 at 2006 gas prices. An additional $100,000 in cost savings was observed by the manufacturer because the conveyor belts were now lasting longer since they were not being constantly heated by the gas pilot lights. This non-energy savings cobenefit was not originally assessed by the IAC team but was caught in the Implementation Report phase of the assessment.

Table 6.8 IAC Recommendations for the Texas Tile Manufacturing LLC Industrial Assessment

Recommended Action	Annual Resource Savings	Annual Cost Savings	Implementation Cost	Payback
Install spark igniters	12,100 MMBtu	$113,000 (+$100,000)	$1300	Less than 1 month
Return condensate	2720 MMBtu	$25,400	$600	1 month
Insulate steam lines	2298 MMBtu	$21460	$6100	4 months
Repair steam line	2260 MMBtu	$21100	$1235	1 month
Relamp with HP sodium lamps	135,800 kWh (250 kVA)	$16,630	$22,400	17 months
Use engineered nozzles	143,000 kWh (250 kVA)	$13,470	$390	1 month
Use synthetic lubricants	104,700 kWh	$8900	0	Immediate
Repair steam leaks	860 MMBtu	$8100	$500	1 month
Lighting management	91,600 kWh	$7800	0	Immediate

Table created using data presented in US DOE IAC database Texas Tile vinyl flooring case study (2006).

REFERENCES AND FURTHER READING

Books and Technical Articles/Reports

Afkhami, B., Akbarian, B., Beheshti, N., Kakaee, A. H., & Shabani, B. (2015). Energy consumption assessment in a cement production plant. *Sustainable Energy Technologies and Assessments*, *10*, 84–89.

Anderson, S. T., & Newell, R. G. (2004). Information programs for technology adoption: the case of energy-efficiency audits. *Resource and Energy Economics*, *26*(1), 27–50.

Aubrecht, G. J., & Aubrecht, I. I. (2006). *Energy: Physical, environmental, and social impact*. Upper Saddle River, NJ: Pearson Prentice Hall.

Blok, K., & Nieuwlaar, E. (2016). *Introduction to energy analysis*. Taylor & Francis.

Ebenhack, B. W., & Martínez, D. M. (2013). *The path to more sustainable energy systems: How do we get there from here?* Momentum Press.

Gielen, D., & Taylor, M. (2007). Modelling industrial energy use: The IEAs energy technology perspectives. *Energy Economics*, *29*(4), 889–912.

He, K., & Wang, L. (2017). A review of energy use and energy-efficient technologies for the iron and steel industry. *Renewable and Sustainable Energy Reviews*, *70*, 1022–1039.

Kesicki, F., & Yanagisawa, A. (2015). Modelling the potential for industrial energy efficiency in IEA's World Energy Outlook. *Energy Efficiency*, *8*(1), 155–169.

Kirova-Yordanova, Z. (2015). Low temperature waste heat in chemical industry: problems and possible ways to recover the heat of condensation. In: *Proceedings of ECOS 2015 — The 28th international conference on efficiency, cost, optimization, simulation and environmental impact of energy systems, Paris, France* (pp. 1–12).

Kluczek, A., & Olszewski, P. (2017). Energy audits in industrial processes. *Journal of Cleaner Production*, *142*, 3437–3453.

Madlool, N. A., Saidur, R., Hossain, M. S., & Rahim, N. A. (2011). A critical review on energy use and savings in the cement industries. *Renewable and Sustainable Energy Reviews*, *15*(4), 2042–2060.

Papapetrou, M., Kosmadakis, G., Cipollina, A., La Commare, U., & Micale, G. (2018). Industrial waste heat: Estimation of the technically available resource in the EU per industrial sector, temperature level and country. *Applied Thermal Engineering*, *138*, 207–216.

Ramirez, C. A., Patel, M., & Blok, K. (2005). The non-energy intensive manufacturing sector: An energy analysis relating to the Netherlands. *Energy*, *30*(5), 749–767.

Salas, D. A., Ramirez, A. D., Rodríguez, C. R., Petroche, D. M., Boero, A. J., & Duque-Rivera, J. (2016). Environmental impacts, life cycle assessment and potential improvement measures for cement production: A literature review. *Journal of Cleaner Production*, *113*, 114–122.

University of York Centre for Industry Education Collaboration, York, UK (2016). *The essential chemical industry — Online*. Available from <http://www.essentialchemicalindustry.org/>.

Worrell, E., Bernstein, L., Roy, J., Price, L., & Harnisch, J. (2009). Industrial energy efficiency and climate change mitigation. *Energy Efficiency*, *2*(2), 109–123.

Energy Information Sources and Technical Reports

Belzer, D.B. (2014). A comprehensive system of energy intensity indicators for the U.S.: Methods, data and key trends. In: *Pacific Northwest National Laboratory Report PNNL-22267*. Available from <https://www.pnnl.gov/main/publications/external/technical_reports/PNNL-22267.pdf>.

Boulamanti, A., & Moya, J.A. (2017). Energy efficiency and GHG emissions: Prospective scenarios for the chemical and petrochemical industry, 2017. In: *European Commission joint research centre report EUR 28471 EN*. Available from <http://publications.jrc.ec.europa.eu/repository/bitstream/JRC105767/kj-na-28471-enn.pdf>.

Bray, E.L. (2012). *United States geologic survey yearbook 2012: Aluminum*. Available from <https://minerals.usgs.gov/minerals/pubs/commodity/aluminum/myb1-2012-alumi.pdf>.

Brueske, S., Sabouni, R., Zach, C., & Andres, H. (2012). U.S. manufacturing energy use and greenhouse gas emissions analysis. In: *US DOE Oak Ridge national laboratory report ORNL/TM-2012/504*. Available from <https://www.energy.gov/sites/prod/files/2013/11/f4/energy_use_and_loss_and_emissions.pdf>.

Central Intelligence Agency (CIA) (2018). *World fact book*. Accessed from <https://www.cia.gov/library/publications/the-world-factbook/>.

Diakosavvas, D. (2017). *Improving energy efficiency in the agro-food chain. Organisation for Economic Co-operation and Development (OECD) Green Growth Studies*. Paris: OECD Publishing. Report COM/TAD/CA/ENV/EPOC(2016)19/FINAL. Available from http://www.oecd.org/publications/improving-energy-efficiency-in-the-agro-food-chain-9789264278530-en.htm.

Food and Agricultural Organization of the United Nations (FAO) (2015a). *FAO yearbook of forest products 2015*. Available from <http://www.fao.org/forestry/statistics/80570/en/>.

Food and Agricultural Organization of the United Nations (FAO) (2015b). *Pulp and paper capacities: Survey 2014−2019*. Available from <http://www.fao.org/forestry/statistics/81757/en/>.

Galitsky, C., Chang, S., Worrell, E., & Masanet, E. (2005). Energy efficiency improvement and cost saving opportunities for the pharmaceutical industry. In: *Lawrence Berkeley Livermore report LBNL-57260*. Available from <https://www.energystar.gov/ia/business/industry/LBNL-57260.pdf>.

Kramer, K.J., Masanet, E., Xu, T., & Worrell, E. (2009). Energy efficiency improvement and cost saving opportunities for the pulp and paper industry. In: *Lawrence Berkeley Livermore report LBNL-2268E*. Available from <https://www.energystar.gov/sites/default/files/buildings/tools/Pulp_and_Paper_Energy_Guide.pdf>.

Menzie, W.D., Barry, J.J., Bleiwas, D.I., Bray, E.L., Goonan, T.G., & Matos, G. (2010). United States geologic survey: The global flow of aluminum from 2006 through 2025. In: *Open-file report 2010−1256*. Available from <https://pubs.usgs.gov/of/2010/1256/pdf/ofr2010-1256.pdf>.

National Academies Press (1999). *Industrial environmental performance metrics: Challenges and opportunities*. Available from <https://www.nap.edu/read/9458/>.

Organisation for Economic Development and Cooperation International Energy Agency (IEA) (2018). *Tracking clean energy progress: Cement*. Available from <https://www.iea.org/tcep/industry/cement/>.

Organisation for Economic Development and Cooperation International Energy Agency (IEA) (2017). *Tracking clean energy progress: Pulp and paper*. Available from <https://www.iea.org/etp/tracking2017/pulpandpaper/>.

US Department of Energy Office of Energy Efficiency and Renewable Energy (EERE) Advanced Manufacturing Office (2017). *Bandwidth study on energy use and potential energy saving opportunities in U.S. aluminum manufacturing*. Available from <https://www.energy.gov/eere/amo/downloads/bandwidth-study-us-aluminum-manufacturing>.

US Department of Energy Office of Energy Efficiency and Renewable Energy (EERE) Industrial Assessment Center (IAC) Database (2006). Vinyl flooring texas tile manufacturing llc, additional savings are a pleasant surprise. In: *Case study report.* Available from <https://iac.university/assessment/AM0517>.

US Department of Energy Office of Energy Efficiency and Renewable Energy (EERE) Industrial Technologies Program (2003). Water Use in Industries of the Future, Available from <http://www.ana.gov.br/Destaque/d179docs/PublicacoesEspecificas/Metalurgia/Steel_water_use.pdf>

US Department of Energy Office of Energy Efficiency and Renewable Energy (EERE) Industrial Technologies Program (2004). *Energy use, loss and opportunities analysis: US manufacturing & mining.* Available from <https://www.energy.gov/eere/amo/downloads/energy-use-loss-and-opportunities-analysis-us-manufacturing-mining>.

US Department of Energy Office of Energy Efficiency and Renewable Energy (EERE) Industrial Technologies Program (2007). *US energy requirements for aluminum manufacturing.* Available from <https://www.energy.gov/sites/prod/files/2013/11/f4/al_theoretical.pdf>.

US Department of Energy Office of Energy Efficiency and Renewable Energy (EERE) Industrial Technologies Program, (2017). *Bandwidth study on energy use and potential energy savings opportunities in U.S. food and beverage manufacturing.* Available from <https://www.energy.gov/sites/prod/files/2017/12/f46/Food_and_beverage_bandwidth_study_2017.pdf>.

US Department of Energy Office of Energy Efficiency and Renewable Energy (EERE) Industrial Technologies Program (2012). *US manufacturing energy use and greenhouse gas emissions analysis.* Available from <https://www.energy.gov/eere/amo/downloads/us-manufacturing-energy-use-and-greenhouse-gas-emissions-analysis>.

US Energy Information Administration (EIA) (2012). *EIA today in energy: Energy needed to produce aluminum.* Available from <https://www.eia.gov/todayinenergy/detail.php?id = 7570>.

US Energy Information Administration (EIA) (2014a). *EIA today in energy: Recycling is the primary energy efficiency technology for aluminum and steel manufacturing.* Available from <https://www.eia.gov/todayinenergy/detail.php?id = 16211>.

US Energy Information Administration (EIA) (2014b). *2014 Manufacturing energy consumption survey.* Available from <https://www.eia.gov/consumption/manufacturing/data/2010/pdf/Table8_1.pdf>.

US Energy Information Administration (EIA) (2014c). *Metal-based durables manufacturers are fueled by electricity and natural gas.* Available from <https://www.eia.gov/todayinenergy/detail.php?id = 17371>.

US Energy Information Administration (EIA) (2016). *International Energy Outlook 2016: With projections to 2040.* Available from <https://www.eia.gov/outlooks/ieo/pdf/0484(2016).pdf>.

US Energy Information Administration (EIA) (2017). *Fuel consumed at refineries.* Available from <https://www.eia.gov/dnav/pet/pet_pnp_capfuel_dcu_nus_a.htm>.

US Energy Information Administration (EIA) (2018). *Energy efficiency, technical change and price responsiveness in non-energy intensive chemicals.* Available from <https://www.eia.gov/analysis/studies/demand/industrial/chemicals/light-ee/>.

US Environmental Protection Agency (EPA) (1995). Profile of the pulp and paper industry. In: *EPA report EPA/310-R-95-015.* Available from <https://archive.epa.gov/compliance/resources/publications/assistance/sectors/web/pdf/print.pdf>.

Worrell, E., Blinde, P., Neelis, M., Blomen, E., & Masanet, E. (2011). Energy efficiency improvement and cost saving opportunities for the US iron and steel industry. An ENERGY STAR (R) guide for energy and plant managers. In: *Lawrence Berkeley national laboratory report LBNL-4779E*. Prepared for the US Environmental Protection Agency. Available from <https://www.energystar.gov/sites/default/files/buildings/tools/Iron_Steel_Guide.pdf>.

Worrell, E., Corsten, M., & Galitsky, C. (2015). Energy efficiency improvement and cost saving opportunities for petroleum refineries. An ENERGY STAR (R) guide for energy and plant managers. In: *US Environmental Protection Agency (EPA) report document number 430-R-15-002*. Available from <https://www.energystar.gov/sites/default/files/tools/ENERGY_STAR_Guide_Petroleum_Refineries_20150330.pdf>.

Worrell, E., & Galitsky, C. (2008). Energy efficiency improvement and cost saving opportunities for cement making (Revision). In: *Lawrence Berkeley National Laboratory. Report number LBNL-54036-Revision*. Available from <https://www.energystar.gov/ia/business/industry/LBNL-54036.pdf>.

Worrell, E., Martin, N., Price, L., Ruth, M., Elliott, N., Shipley, A., et al. (2004). Emerging energy-efficient technologies for industry. In: *Lawrence Berkeley National Laboratory and American Council for an Energy Efficient Economy (ACEEE) joint report. Report number LBNL-57572*. Available from <https://ies.lbl.gov/publications/emerging-energy-efficient>.

Worrell, E., Phylipsen, D., Einstein, D., & Martin, N. (2000). Energy use and energy intensity of the US chemical industry. In: *Lawrence Berkeley National Laboratory report LBNL-44312*. Available from <https://www.energystar.gov/sites/default/files/buildings/tools/industrial_LBNL-44314.pdf>.

Worrell, E., Price, L., Neelis, M., Galitsky, C., & Zhou, N. (2007). World best practice energy intensity values for selected industrial sectors. In: *LBNL report number LBNL-62806-REV2*. Available online from <https://eaei.lbl.gov/sites/all/files/industrial_best_practice_en.pdf>.

Other Online Resources

American Chemical Society (ACS) (2009). Sustainable U.S. manufacturing in the chemical and allied industries. In: *ACS Vision 2020 workshop*. Available from <http://www.acs.org/smrt>.

Edwards, P. (2017). Global cement top 100 report 2017–2018. *Global Cement Magazine*. Available from <http://www.globalcement.com/magazine/articles/1054-global-cement-top-100-report-2017-2018>.

Hannah, M., Harmon, J., & Petersen, D. (2011). *Estimating payback for energy efficiency*. Energize Ohio. Ohio State University Extension. Available from <https://energizeohio.osu.edu/sites/energizeohio/files/d6/files/imce/PM-2089S.pdf>.

International Aluminum Institute (2015). *Industry structure*. Available from <http://recycling.world-aluminium.org/review/industry-structure/>.

International Aluminum Institute (2018). *World aluminum primary aluminum production*. <http://www.world-aluminium.org/statistics/primary-aluminium-smelting-power-consumption/#data>.

Portland Cement Association (2018). *Cement: The foundation of civilization*. Available from <https://www.cement.org/civilization>.

Thermal Energy International (2017). *Improving energy efficiency in the pulp and paper industry*. Available from <http://www.thermalenergy.com/blog/improving-energy-efficiency-in-the-pulp-and-paper-industry>.

UC Rusal (2017). *How the world aluminum market works*. Available from <https://www.aluminiumleader.com/economics/world_market/>.

World Steel Association. (2008). *Sustainability report of the world steel industry*. Brussels, Belgium: World Steel Association, 34pp.

World Steel Association (2018). *World steel in figures 2018*. Brussels, Belgium: World Steel Association. Available from <https://www.worldsteel.org/media-centre/press-releases/2018/world-steel-in-figures-2018.html>.

Transportation sector energy efficiency

7

CHAPTER OUTLINE

The transportation sector contains those elements of society that allow for the movement of goods and people across large tracts of land and over wide expanses of water. In this chapter, we review:

- consumption in common transportation modes;
- energy processes and efficiencies within passenger vehicles and freight road transport.

7.1 TRANSPORTATION SECTOR OVERVIEW

Transportation is the linchpin of local, regional, and global economies. Goods are produced across the globe and shipped to consumers, often traveling thousands of kilometers in the process. Moreover, personal mobility now defines affluent society. The ability to travel quickly from home to office or factory, to the school or church, and to the football field or market is very much tied to access to stable and secure roadways and vehicles that move us about. The ability to fly across oceans in just a few hours has made the world a much smaller place than ever before in human history, allowing us to move distances and at speeds impossible using our own muscles. However, these abilities have also come at a

Energy Efficiency. DOI: https://doi.org/10.1016/B978-0-12-812111-5.00007-X

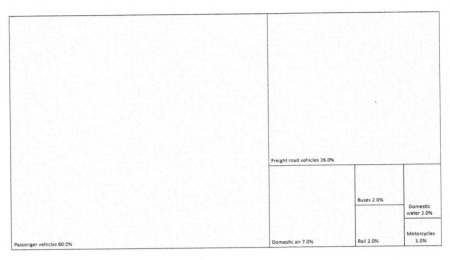

FIGURE 7.1

Distribution of transport energy by mode in Organisation for Economic Cooperation and Development member countries (i.e., Developed Nations), 2016. All road vehicle classifications account for 89% of the total energy consumption by modes using commercial energy.

Figure created with data presented in IEA's Energy Efficiency Indicators Highlights (2018).

significant energy requirement that continues to grow as more people gain access to more diverse, energy-intensive transportation options. Indeed, as described in Chapter 6, Industrial Sector Energy Efficiency, the global transportation sector is fast approaching the consumption of 121 EJ (115 Quads) of energy annually, mostly from the fossil fuels.

The International Energy Agency (IEA) defines the transportation sector, or transport sector, as "all transport modes using commercial energy, independently of the sector where the activity occurs." This includes (1) road vehicles, comprising passenger cars, freight road vehicles, buses, and motorcycles; (2) air; (3) rail; and (4) water. Countries in the Organisation for Economic Cooperation and Development (OECD), for which the IEA tracks energy consumption, use more energy for transport than they do for any other end-use activity. As recently as 2016, 89% of all commercial transport energy was consumed by road vehicles, with the distribution depicted in Fig. 7.1.[1]

Each transport mode requires a slightly different commercial energy source/ fuel to match the specification of their propulsion system (usually an engine

---

[1]IEA does not include international travel or pipeline transport in this accounting. Other modes not considered include walking and bicycling, which do not require commercially sold energy to operate.

and/or a motor), which, according to the US Energy Information Administration (EIA), includes:

- gasoline,
- diesel and other distillate fuels,
- jet fuel,
- residual fuel oil,
- propane,
- ethanol (fuel alcohol) and biodiesel,
- natural gas, and
- electricity.

Some of these energy sources and fuels are used in particular propulsion systems, while others can be used to power a diversity of systems as depicted in Fig. 7.2.

The most noteworthy fuels with respect to passenger and freight road transport globally are the oil products: gasoline and diesel. According to the IEA, motor gasoline is the dominant fuel for passenger cars, used in 82% of all cars, light trucks, and sport utility vehicles (SUVs) in the OECD. This is

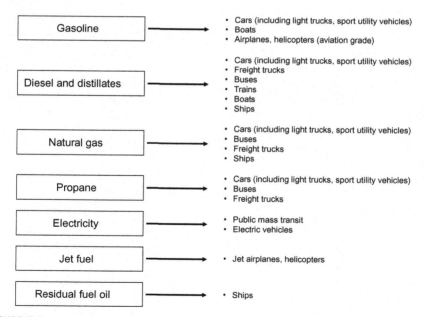

FIGURE 7.2

Transportation fuels and their common applications. Note: biofuels are not noted as a separate source/fuel because they are typically blended in with gasoline and/or diesel.

Figure created with information presented in EIA's Energy Explained: Energy Use for Transportation (2018).

down from 90% in 2000, the reason for the decline coming from a marked increase in diesel car usage, primarily in Europe. Diesel remains the dominant fuel for freight road transport at over 90%. In either scenario, petroleum is the single largest primary energy source for global transportation. The IEA, for example, reports that in 2016, oil products represented 92% of transportation sector energy use for the entire world (OECD plus non-OECD). Consumption data for some selected countries, including primary mode, are provided in Table 7.1.

The key energy efficiency indicators tracked with respect to transportation sector energy consumption are *passenger energy transport intensity* and *freight energy transport intensity*. Passenger energy transport intensity is defined as "the amount of energy used to move one passenger over a distance of one kilometer" and is measured in units of megajoules per person-kilometer (MJ/pkm). Freight energy transport intensity is defined as "the amount of energy used to move a tonne of freight over a distance of one kilometer" and is measured in units of megajoules per tonne-kilometer (MJ/tkm).

Fig. 7.3 is derived from a recent IEA report, which provides trends in *passenger* transport for a recent trend in intensity from select countries, comparing the

Table 7.1 Transportation Consumption for Select Countries in Units of Million Tonnes of Oil Equivalent (MTOE), Including Percent of Consumption Energy From Oil Products, and Primary Mode of Transport, Including Percent Share of All Transport Modes Used by That Country

Country	Transportation Consumption (MTOE)	% From Oil Products	Primary Mode, and Percent of All Transport
United States	622.0	91%	Road (85.0%)
China	296.0	90%	Road (82.1%)
Germany	56.8	93%	Road (95.2%)
Denmark	4.2	93%	Road (91.6%)
Norway	4.8	89%	Road (71.5%)
Iceland	0.3	94%	Road (93.3%)
United Kingdom	41.0	97%	Road (94.1%)
Canada	61.1	90%	Road (76.3%)
Spain	30.6	94%	Road (89.5%)
Australia	32.9	97%	Road (82.6%)
Switzerland	5.7	93%	Road (93.5%)
France	43.8	91%	Road (94.5%)
India	90.0	95%	Road (91.4%)

Transport modes are exclusive to those that require commercial fuel to operate. Secondary mode of transport is usually domestic air or domestic navigation.
Table created with data presented in IEA's Atlas of Energy: Energy Balance Flows (2016).

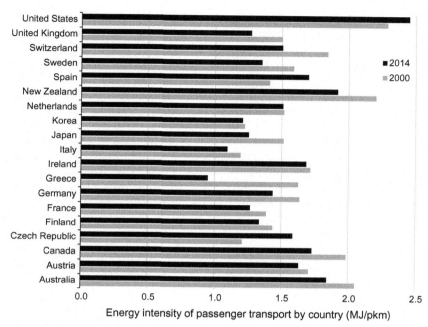

FIGURE 7.3

Passenger transport energy intensity in OECD member countries, c. 2014, in units of MJ/pkm. Countries listed represent roughly 85% of total consumption in OECD. *MJ/pkm*, megajoules per passenger-kilometer; *OECD*, Organisation for Economic Cooperation and Development.

Figure created with data presented in IEA's Energy Efficiency Indicators Database (2017).

years 2000 and 2014. According to the IEA, the intensity varies from country to country based on a number of factors, including the following:

- Share of transportation mode, i.e., road transport versus air, rail, water, etc.
- Vehicle mix in the transportation stock, i.e., passenger car versus bus, motorcycle, etc.
- Average occupancy, i.e., the average number of passengers per vehicle.

Notice that, of the countries listed in Fig. 7.3, the United States, Canada, Australia, and New Zealand all have higher passenger transport intensities than the rest of the OECD, attributed to the large use of passenger cars, particularly light trucks and SUVs, as well as higher usage of domestic air, and also higher per capita distances traveled (all are geographically larger than most other countries). Also, notice in Fig. 7.3 that almost all countries listed have reduced intensity in that 14-year time frame. A reason for much of the reduction is attributed to increased passenger efficiency related to a shift to diesel passenger car usage in Europe. In other cases, reductions are attributed to modal shifts (more use of trains and/or buses). The United States, Spain, and the Czech Republic have all increased in intensity in this time frame, likely related to reduced average occupancy, increased use of domestic air transport, and increased usage of SUVs. Of course, much of this increased intensity is also a function of low fuel prices.

Finally, it is notable that since the IEA started tracking sectoral energy consumption in 1973, the percent of the energy consumed for transport has been steadily increasing globally. Whereas in 1973, 23% of total final global consumption came from transport energy consumption, in 2016, that percentage rose to 28.8%. In the OECD only, the percentage rose from 24.7% of total final consumption to 33.7% of total. Just focusing on China, since 2000, transportation energy consumption has increased from 84 to 296 MTOE, a 350% increase. As countries continue to develop, the need for transportation energy will continue to rise. While it has always been important to increase transport efficiency, these consumption trends indicate that an even greater effort will be needed in order to control global transportation energy consumption.

7.2 MODES OF TRANSPORT IN THE SECTOR

Based on the sector overview presented in Section 7.1, the most important energy-intensive transportation modes globally in descending order are

- passenger vehicles;
- freight road vehicles;
- air;
- buses, rail, and water; and
- motorcycles.

Of these transport modes, passenger vehicles and freight road vehicles make up 86% of total transport energy consumption. As such, we will focus much of our attention on motor and freight vehicles, with an additional smaller focus on air, which is the next most energy consumptive, comprising 7% of total transport energy consumption.

7.2.1 MOTOR VEHICLES AND AUTOMOTIVE TRANSPORTATION

Motor vehicles, also known as automobiles, are vehicles driven by a motor, typically by an *internal combustion engine* (ICE) or equivalent propulsion system, used by vehicle operators to transport goods and people on roads and highways. As per Fay and Golomb, governments set regulatory objectives that automobile manufacturers must follow, not only for manufacturing but also for operation and maintenance of motor vehicles. These include

- vehicle and passenger safety,
- operator competence,
- owner fiscal liability,
- control of exhaust and evaporative emissions, and
- vehicle fuel economy.

The objective most relevant to energy consumption is vehicle fuel economy, or how far a vehicle can travel per unit of fuel; however, control of exhaust and evaporative emissions and vehicle and passenger safety also contribute in indirect ways to energy consumption. In addition, energy suppliers are expected to comply with the most current set of standards for refining and distributing the fuels used to power motor vehicles.

As such, vehicles are classified on many different bases, including primary fuel used, load type, number of wheels, number of axles, body type, transmission, and emissions. Since vehicles are intricately tied to fuel and fuel costs, the focus tends to be on fuel efficiency, thus motor vehicles tend to be evaluated mostly on the basis of fuel used. Using the US transportation sector as an example, the chief motor vehicles categorized by fuel include

- gasoline (or petrol) vehicles, which includes ethanol blends,
- diesel vehicles, which includes biodiesel blends,
- compressed natural gas or liquid petroleum gas vehicles,
- electric vehicles (EVs),
- hybrid gasoline/electric vehicles (HEV), and
- hydrogen fuel cell vehicles.

Because motor vehicles also include freight transport, it is important to also specify vehicle class by load or weight. Different countries use slightly different metrics for classification, and even within a country, different classifications may arise out of different government agencies. For example, in the United States, three agencies that define vehicle weight differently are the Federal Highway Administration (FHWA), the US Environmental Protection Agency (EPA), and the US Census Bureau. Each agency has devised their classification scale based on how they impact highways and roads to determine safe weight limits (FHWA), how they emit pollutants (EPA), and how they are used by the population, in general (US Census). In the United States, classifications are typically made by gross vehicle weight ratings, as shown in Table 7.2 using FHWA and US Census classifications.

Examples of vehicles by weight class using the US FHWA system include the following:

- Class 1: Passenger car, pickup truck, minivan, SUV
- Class 2: Full-size pickup truck, minibus, step van, utility van

Table 7.2 Vehicle Weight Classes and Categories Used by the US Federal Highway Administration and the US Census Bureau

| GVWR (lbs) | US Federal Highway Administration | | US Census Bureau |
	Vehicle Class	GVWR Category	VIUS Classes
<6000	Class 1: <6000 lbs	Light duty <10,000 lbs	Light duty <10,000 lbs
10,000	Class 2: 6001 − 10,000 lbs		
14,000	Class 3: 10,001 − 14,000 lbs	Medium duty 10,001 − 26,000 lbs	Medium duty 10,001 − 19,500 lbs
16,000	Class 4: 14,001 − 16,000 lbs		
19,500	Class 5: 16,001 − 19,500 lbs		
26,000	Class 6: 19,501 − 26,000 lbs		Light−heavy duty 19,001 − 26,000 lbs
33,000	Class 7: 26,001 − 33,000 lbs	Heavy duty >26,001 lbs	Heavy duty >26,001 lbs
>33,000	Class 8: >33,001 lbs		

Classes are based on gross vehicle weight ratings (GVWR), is the maximum operating weight, which includes passengers, cargo, and fluids. US census classes based on vehicle inventory and use survey (VIUS). GVWR, Gross vehicle weight rating.
Table created with source data presented in the Alternative Fuels Data Center: Maps and Data (2012).

- Class 3: City delivery truck, walk-in van/truck
- Class 4: Conventional van, landscape utility van, large walk-in van/truck
- Class 5: Bucket truck
- Class 6: Beverage truck, school bus, stake body
- Class 7: City transit bus, furniture truck, home fuel truck, tow truck
- Class 8: Cement mixer, fire truck, heavy semitractor, tour bus

According to the Oak Ridge National Laboratory (ORNL) Transportation Energy Data Book (TEDB), the total amount of registered cars and trucks globally in 2016 exceeded 1.3 billion (1300 million), led by the United States and China with 270 million and 194 million cars and trucks, respectively. In terms of new vehicle production, the TEDB reports that global production of cars and trucks (including buses) in 2016 exceeded 56.5 million and 37.48 million, respectively. Assuming these new vehicles were all registered, the 2016 value represents nearly 7% of the total registered vehicle stock globally. Figs. 7.4 and 7.5 use TEDB data to show the rapid increase in vehicle production between 1986 and 2016 for the world, with the largest growth in recent production occurring in China, India, Indonesia, South Korea, Thailand, and the Czech Republic. China, as with every other economic sector since about 2000, is by far the largest growing producer of cars and trucks.

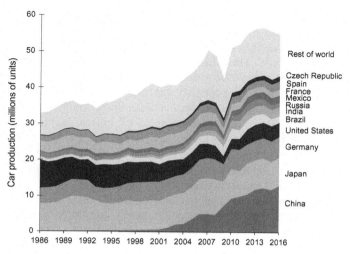

FIGURE 7.4

World car production from 1986 to 2016 (i.e., cars are defined as passenger vehicles not including light trucks/SUVs for proper global comparison). *SUV,* Sport utility vehicle.

Figure created with source data presented in ORNL's Transportation Energy Databook, Edition 37 (2018).

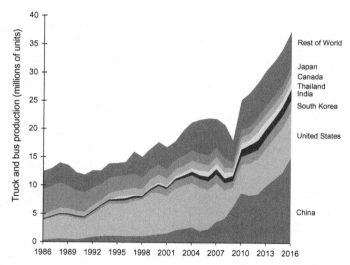

FIGURE 7.5

World truck and bus production from 1986 to 2016 (i.e., trucks include light trucks/SUVs for proper global comparison). *SUV,* Sport utility vehicle.

Figure created with source data presented in ORNL's Transportation Energy Databook, Edition 37 (2018).

In terms of per capita vehicle stocks, the TEDB reports that the United States remains the most reliant on motor vehicles with a 2016 value of 831.9 vehicles per 1000 people. For comparison, Western Europe's per capita value was 606.0 vehicles per 1000 people. The TEDB also notes that China's vehicle stock rose sharply between 2006 and 2016, from 26.6 vehicles per 1000 people to 141.2 vehicles per 1000 in that time frame. If that rate were to continue over the next 10 years (2016−26), China's per capita vehicle stock would rise to nearly 750 vehicles per 1000 people. With a population of 1.4 billion people, that per capita value would translate to having 1 billion vehicles in the country. Even if the per capita value only rises to half that value in 2026 (i.e., 375 vehicles per 1000), half a billion new vehicles on the road in roughly 10 years will likely lead to an entirely different kind of transportation energy landscape.

Given these recent trends in global vehicle stock, the US EIA projects that over the next 25 years, OECD transportation energy consumption will remain relatively flat, while non-OECD consumption will continue to grow and surpass OECD in the early 2020s, with total consumption in the transportation sector growing through at least 2040. Both the EIA and IEA agree that the impact of more vehicles on roads globally, but especially in non-OECD countries, ties directly to three important features of future road transportation energy consumption: fuel economy, fuel choice, and transport mode choice. Assuming the current transport mode choice of a personal vehicle remains highly valued by consumers globally, energy analysts believe that changes in vehicle size, vehicle fuel economy, and/or fuel choice will affect future consumption patterns more quickly in countries with millions of first-time vehicle purchasers than in countries with already existing large vehicle stocks.

As long as fuel prices do not rise dramatically in a short time period, the largest factor that will affect total energy consumption from motor vehicles going forward is the fuel economy standards set in place by governments across the world.[2] Most countries in the OECD and non-OECD have enacted some type of light-duty and heavy-duty vehicle fuel economy standards, which regulates the fuel efficiency of vehicles that vehicle manufacturers can produce and sell. Some official standards are more stringent than others, with OECD countries having stricter regulations, and rapidly developing non-OECD countries tending to have weaker or no fuel economy standards (policy and regulations will be described more completely in Chapter 9: Policy instruments to foster energy efficiency). Most are tied to reducing petroleum fuel import dependence to control gasoline and diesel fuel price volatility. Moreover, the more stringent these standards become, the less likely manufacturers will be able to rely on the traditional ICE to satisfy them, due to technical constraints. As such, these standards have also

---

[2]Historically, consumer choices are not centered on fuel economy but rather personal preferences. This has been possible, especially in countries like the United States, due to historically low transportation energy prices relative to other factors. Only during times of very high fuel prices has there been a consumer focus on fuel economy.

encouraged alternative power plants, especially HEVs and full EVs, where a much higher equivalent fuel economy can be achieved.

According to Sperling and Lutsey, thus far, the efficiency improvements in conventional ICE vehicles chiefly have come from

- more efficient combustion, including gasoline direct injection, homogeneous-charge compression ignition in diesel, and variable valve timing;
- turbocharging,
- smarter cooling systems;
- reduced engine friction;
- more efficient transmissions;
- lighter weight materials; and
- better aerodynamic design.

The degree to which additional advancements in these areas can significantly improve overall fuel efficiency has also been achieved through the use of hybrid electric drives. By coupling ICEs to electric motors, smaller engines are required to move the vehicle, which reduces fuel consumption.

All of these technological changes have helped improve real vehicle fuel economy, especially in the light-duty vehicle class. Fig. 7.6 depicts the US EPA's adjusted US fuel economy for different light-duty vehicles in units of miles per gallon from 1975 to 2017. After the oil embargo of 1973−74, constrained fuel supplies in the United States resulted in massive increases in the price of gasoline

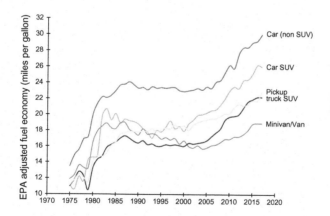

FIGURE 7.6

Adjusted fuel economy for US passenger vehicles between 1975 and 2017, in units of miles per gallon. Adjusted fuel economy differs from the US Corporate Average Fuel Economy standard by accounting for more realistic driving conditions than those used by manufacturers to comply with the standard. Car SUVs are built on car frames and truck SUVs are built on truck frames, which affects fuel economy. *SUV,* Sport utility vehicle.

Figure created with data presented in EPA's Light-duty Vehicle CO$_2$ and Fuel Economy Trends (2018).

at the pump. This incentivized the US government to introduce the Corporate Average Fuel Economy standard, which required automobile makers to increase vehicle fuel efficiency.

As can be seen in Fig. 7.6, these standards resulted in a significant increase in vehicle fleet fuel economy between 1975 and 1985. Not surprisingly, increases in fuel economy stagnated between 1985 and 2005 as transportation energy prices retreated. Then, as fuel prices once again rapidly began to increase around 2005, fuel economy also began to rise, which likely coincided with wider adoption of smaller vehicles, diesel, and HEVs. Clearly, one of the primary motivations for innovation to improve efficiency is to control fuel price volatility by reducing energy demand. Putting Fig. 7.6 into the context of fuel saved, it is helpful to compare the amount of fuel used by passenger vehicles and by large trucks (tractor/trailer combinations) to the amount of fuel that would have been required, based on the vehicle miles traveled, if efficiency (or the surrogate, fuel economy in mpg) had not improved after 1980.

Fig. 7.7 depicts the two fuel scenarios for passenger cars. The lower (dark, solid) curve represents the amount of fuel actually consumed by cars in the United States from 1980 to 2016, while the upper, dashed curve shows how much fuel would have been necessary for cars to travel the same mileage as was observed each year, but without the average fuel economy improvements which actually occurred during those 35 years. There are some irregularities in the data, related to changes in data handling: specifically changes in counting vehicles as

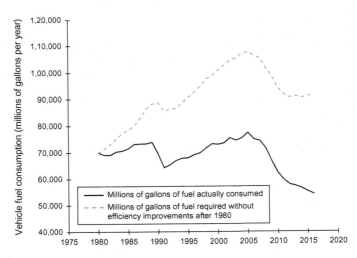

FIGURE 7.7

A comparison of passenger vehicle fuel efficiency in the United States to the amount that would have been required if efficiency had not improved after 1980 passenger vehicle fuel consumption with and without efficiency improvements (millions of gallons per year).

Figure created using data presented in ORNL's Transportation Energy Databook, Edition 37 (2018).

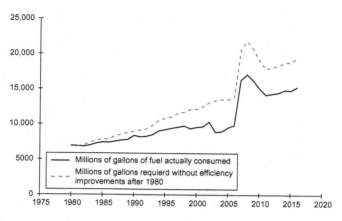

FIGURE 7.8

A comparison of the transportation fuel used by large trucks in the United States, to the amount that would have been required if efficiency had not improved after 1980 truck fuel consumption with and without efficiency improvements after 1980.

Figure created using data presented in ORNL's Transportation Energy Databook, Edition 37 (2018).

cars in 1985 and then another change in data methodology in 2009. However, the overall trends remain quite clear, demonstrating a growing savings in energy use, in spite of generally increasing mileage driven.

As with the previous figure, Fig. 7.8 shows the fuel savings over time: in this case, as the efficiency of large trucks improved. Again, the lower, solid curve represents actual fuel consumption, while the upper, dashed curve shows the fuel that would have been consumed if the truck efficiencies had remained constant at 1980 levels. In the case of trucks, the increased mileage driven was more steadily increasing through the 36-year time period considered. The overall fuel consumption, then, does continue to rise, but not as rapidly as it would have without the efficiency gains. By the end of the period, almost 5 billion gallons of diesel were being conserved annually.

In spite of the dramatically greater fuel consumption per mile by trucks, there are so many more cars, driving so many miles, that the total savings from increased car efficiencies are much larger than the savings from improved efficiency in the trucking fleet. The combined total over the 31-year time period considered is 900 billion gallons: more than 20 billion barrels. This is more than even a very large single oil field—and a little less than one year's global consumption of petroleum.

7.2.2 AIR TRANSPORT

Air transport, which represents the next most substantial energy-consuming transport sector, includes passenger and freight airplanes, that is, aircraft configured for transporting passengers, freight, or mail. According to the International Air Transport Association (IATA), in 2017, airlines carried 4.1 billion passengers globally. This value increased by 7.3% over 2016, which represented an additional 280 million trips by air between 2016 and 2017. In addition, as with many of the energy and transport-related statistics in recent times, airlines in the Asia-Pacific region carried the largest number of passengers. According to IATA statistics, the market share of passengers increased from 2016 to 2017 by region is as follows:

1. Asia-Pacific, 36.3%; 1.5 billion passengers (10.6% increase from 2016).
2. Europe, 26.3%; 1.1 billion passengers (8.2% increase).
3. North America, 23%; 941.8 million passengers (3.2% increase).
4. Latin America, 7%; 286.1 million (4.1% increase).
5. Middle East, 5.3%; 216.1 million (4.6% increase).
6. Africa, 2.2%; 88.5 million (6.6% increase).

In addition, it is noteworthy that the aggregated global number of 4.1 billion has doubled since 2005, and by 2036, IATA anticipates that airlines will carry nearly 8 billion passengers globally.

In terms of cargo, the global cargo market expanded 9% in freight and mail tonne-kilometers (tkm). According to IATA, this expansion outstripped a capacity increase of 5.3% with 2017 cargo carrier rankings as follows:

1. Federal Express (16.9 billion tkm)
2. Emirates (12.7 billion tkm)
3. United Parcel Service (11.9 billion tkm)
4. Qatar Airways (11 billion tkm)
5. Cathay Pacific Airways (10.8 billion tkm)

Since airfares are connected to carrier fuel costs (among others), there is an incentive for the industry to reduce fuel consumption as much as possible, both in their aircraft and in their airline infrastructure. According to a 2012 World Bank report, energy efficiency in the air transport sector has come from technology improvements in airframe and engine design, air traffic control, and airport operation. This has resulted in the current stock of aircraft being 80% more fuel efficient than their 1960s counterparts. According to the Intergovernmental Panel on Climate Change, fuel reductions since the 1970s have also been greatly aided by the development and deployment of automatic flight management systems. As evidenced by the improvements that have been achieved in air transport with respect to reducing fuel consumption, further improvements will be similar in scope to improving vehicle efficiency, focusing on technology (e.g., aircraft designs, lightweight materials, new engine advances), operations, and infrastructure.

7.2.3 PASSENGER AND FREIGHT ENERGY INTENSITIES BY MODE

As described in Section 7.1, the key energy efficiency indicators tracked by the IEA with respect to transportation sector energy consumption are passenger

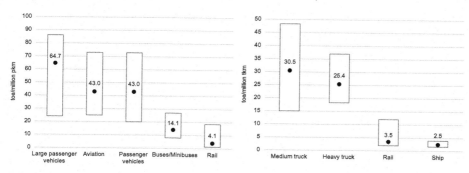

FIGURE 7.9

Global passenger and freight energy intensity by mode in units of toe/million pkm and toe/ million tkm. The bars represent tabulated ranges and black circles represent average intensity for that particular mode. *toe/million pkm*, tonnes of oil equivalent per million passenger kilometers; *toe/million tkm*, tonnes of oil equivalent per million freight kilometers.

Figure created with information and imagery presented in IEA's The Future of Rail (2018), with modification.

energy transport intensity and freight energy transport intensity. It was also mentioned that common units of measure are megajoules per person-kilometer (MJ/ pkm) for passenger intensity and megajoules per tonne-kilometer (MJ/tkm) for freight intensity. When looking at the data from the perspective of global intensity numbers, it is also common to report them in units of tonnes of oil equivalent per million kilometers (toe/million pkm or toe/million tkm). A recent IEA study reports ranges for energy intensity for common modes of passenger and freight energy transport, which is reproduced in Fig. 7.9.

Based on the data presented in Fig. 7.9, the most utilized modes of transport (i.e., passenger vehicles, aviation, and trucks) tend to have the highest average energy intensities and widest range in intensities. The transport infrastructure of many countries has been designed with road vehicles in mind, and it likely will be challenging to replace the infrastructure to suit less energy-intensive modes, such as rail, in many locations. The emphasis going forward will likely have to be reducing the average intensity and intensity ranges in road transport through better and/or alternative designs and incentivizing consumers to choose the least intensive of the transport options.

7.3 ENERGY EFFICIENCY POTENTIAL IN MOTOR VEHICLES

Regardless of fuel type, all vehicles are constructed with a few basic systems important to fuel consumption: power plant, transmission, frame/chassis/body, and accessory/convenience components. The power plant is the primary driver of a motor vehicle and is by far the efficiency-limiting component. The two most common types in passenger and freight road vehicles are spark ignition and

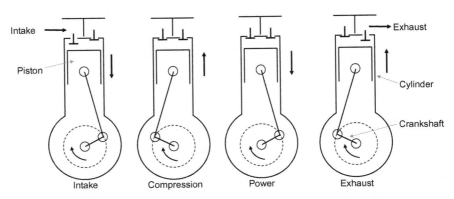

FIGURE 7.10

Depiction of a four-stroke, piston internal combustion engine. The four strokes are intake, compression, power, and exhaust. The fuel/air mixtures enter through an intake valve and spent gases exit through an exhaust valve. (Otto cycle requires a spark during the compression/power strokes.) The cycle repeats as long as the engine is in operation and has sufficient fuel.

Figure created using a composite of depictions presented in Masters and Ela (2008) and Fay and Golomb (2012).

compression ignition ICEs. As was mentioned in Section 7.1, most passenger vehicles sold in the world are gasoline-powered ICEs, which operate according to the Otto engine cycle, typically a four-stroke, *spark-ignited*, piston engine. Freight transport mostly uses the Diesel engine cycle, which is similar to the Otto cycle, however, is a *compression-ignited* cycle. A generic four-stroke cycle is depicted in Fig. 7.10).

In the classic Otto cycle, the four strokes are intake, compression, combustion, and exhaust. There is a series of cylinders and pistons in the engine, designed to move up and down, controlling how a fuel/air mixture passes through, connected to a crankshaft that transfers the chemical energy to the drivetrain and moves the wheels of the vehicle. According to Masters and Ela, the intake step allows the mixture to enter through an intake valve, while the piston is driven down. In the compression step, the piston is driven back up, which compresses the fuel mixture. A spark plug is then used to ignite the mixture. In the power stroke, the temperature and pressure rapidly rise, creating an explosion that drives the piston down and turns the crankshaft, delivering power to the drivetrain. Finally, in the exhaust stroke, the spent gases are released through the exhaust valve. As we said earlier, the diesel cycle operates similarly, but the engine compresses the air more than in the Otto cycle, and introduces the fuel at the top of the compression stroke, which allows the fuel to self-ignite, without needing a spark.

As discussed in Chapter 4, Energy Flow Analyses and Efficiency Indicators, engine efficiency is equivalent to thermal efficiency, dealing with how much chemical energy in gasoline or diesel is converted into work needed to move the

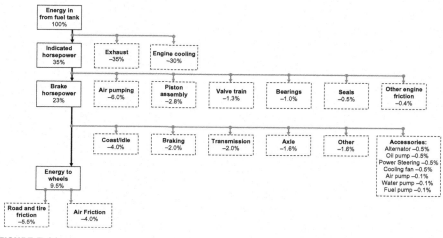

FIGURE 7.11

A schematic accounting of all the energy lost in an ICE passenger vehicle while in operation. Negative signs represent energy lost due to conversion process step. Indicated horsepower is the effective power at the flywheel and brake horsepower is the useful horsepower when accounting for typical engine losses. Note that speed and weight affect this accounting.

Figure based on a depiction presented in Fenske (2009), with modification.

pistons in the engine that then move the car. They operate under the same efficiency limitations as other thermal engines. In Chapter 4, Energy Flow Analyses and Efficiency Indicators, we noted that reported thermal efficiencies for standard gasoline ICEs can range from 20% up to 38%, with an average of roughly 25%, while diesel ICEs average closer to 35% (in part due to higher compression ratios). As per Romer, most of the energy in ICEs is thermally dissipated as exhaust gases and to heating up the cooler surrounding air around the engine. The other component, mechanical efficiency, depends on the mass of the vehicle (including passengers and cargo) and the vehicle's cruising velocity but is an additional source of energy loss due to friction losses in the engine and additional losses from the transmission/drive train, air drag, and rolling resistance of tires. For passenger vehicles, the sum of all possible losses typically results in only about 10% of the original energy content of the fuel actually being utilized to move the vehicle, with a detailed accounting of all possible energy losses provided in Fig. 7.11, as per an Argonne National Laboratory analysis, assuming a 35% thermal efficiency. (Freight vehicles losses are typically even greater.)

Though substantial progress has been made in the last 10 years to improve ICE vehicular fuel economy, Fig. 7.11 demonstrates that the limitations of the ICE in the face of more stringent fuel economy standards globally will most likely contribute to more investment from manufacturers to produce more vehicles with alternative power plants, such as in the aforementioned HEV, including plugin hybrids (PHEV), and EV systems. An electric motor is much more efficient at converting the electrical energy stored in a battery into useful automotive work, at a rate of roughly 90%. In HEV systems, this allows for the use of much smaller ICEs and that weight decrease provides greater overall efficiency. Regenerative braking (described later) is an additional feature that benefits from the use of battery/electric motor systems. This feature is especially beneficial in city driving, where braking is frequent. Differences in fuel economy for a handful of 2018 models, as reported by the US Department of Energy (DOE) and US Environmental Protection Agency (EPA), are shown in Table 7.3.

Finally, because fuel economy is defined by federal agencies in terms of petroleum consumption, electric and plug-in hybrid vehicles are technically more fuel efficient based on an equivalent fuel economy metric, measured in units of miles per gallon equivalent (mpg_e). For example, because the battery pack of a PHEV has been charged off-road using electricity, and because PHEVs can operate without petroleum fuel for a dedicated driving range before the ICE needs to

Table 7.3 Fuel Economy for a Handful of 2018 Passenger Vehicles That Have Conventional Gasoline and Hybrid Gasoline Models Sold in the United States in MPG

2018 Vehicle Make and Model[a]	Conventional Gasoline Powertrain City/ Highway MPG	Hybrid Gasoline Powertrain City/Highway MPG	Hybrid to Gasoline % Increase City	Hybrid to Gasoline % Increase Highway
Chevrolet Malibu	22/32	49/43	223%	134%
Ford Fusion	23/34	43/41	187%	121%
Honda Accord	29/35	47/47	162%	134%
Hyundai Sonata	26/36	40/46	154%	128%
Toyota Camry	28/39	51/53	182%	136%
Toyota Highlander (SUV)	19/26	29/27	153%	104%

Note that hybrid fuel economy in city driving is often much greater than conventional gasoline engines due to the benefit of regenerative braking technology and an ability to rely more on the electric motor. This is reduced in highway conditions, due to the negating effects of increased air drag at highway speed.
MPG, miles per gallon; SUV, sport utility vehicle.
[a]*For reference 2008 city/highway fuel economy for the conventional gasoline Malibu, Fusion, Accord, Sonata, Camry, and Highlander were 22/32, 20/28, 21/31, 21/30, 21/31, and 17/23 MPG, respectively. The largest differences between 2008 and 2018 conventional gasoline model years are that in 2018 there is much more use of turbocharging and six-speed or higher transmissions in drivetrains. Table created using data presented in US DOE/EPA's fueleconomy.gov: Find and Compare Cars (2018).*

be engaged, its mpg$_e$ can be much greater than a conventional gasoline motor vehicle. For an EV, the comparison is different since no petroleum fuel is ever used; however, there is an equivalent conversion between the energy density of the battery pack, measured in kWh and the equivalent energy content of petroleum fuel. The largest difference between the two is that the energy content of a battery pack is much less than a standard gasoline tank, thus driving range in an EV can be significantly shorter than a conventional gasoline-powered vehicle or PHEV.

For example, by comparing the basic energy input of a gasoline-powered vehicle to an EV, we can calculate that 1 gal of gasoline, on average, contains just under 34 kWh/gal. An EV, such as the Nissan Leaf, has a cruising range of about 151 mi, which is directly related to the size of the battery pack in the vehicle. The Leaf requires 7.5 hours to charge at a power draw of 6.6 kW, which is equivalent to 49.5 kWh electrical energy. Dividing the 151-mi range by this 49.5 kWh is equal to an electrical fuel economy of 3.05 mi/kWh. Therefore, the Leaf can be said to have a fuel economy equivalent to

$$3.05 \text{ mi/kWh} \times 34 \text{ kWh/gal} = 106.4 \text{ mpg}_e$$

This calculated value compares to 112 mpg$_e$ reported in the Nissan corporation literature.

This calculation shows the electric-powered vehicle to be much more efficient than comparable gasoline (or even diesel-fueled) vehicles. Recall, though, that the inefficiencies of electric power systems often occur at the step of generating electricity, not in the EV power plant. If a coal-fired power plant is employed, the net system efficiency is reduced by the roughly 70% typical inefficiency at the electricity generator. This would make the EV system efficiency approximately 33 mpg, based on the original fuel input. This would still represent a relatively efficient vehicle, but not as dramatic as the mpg$_e$ may suggest. However, if the electric power originates from a more efficient system, the efficiency benefits of electric propulsion can be fully realized.

Fig. 7.12 illustrates a comparison of passenger vehicle efficiencies, both of the vehicle, and of the vehicle plus original source of energy. The graphics are meant to show the meaning of the term "mpg$_e$" used to characterize EVs. Since it is based on the energy converted within the vehicle, it does successfully show the higher efficiencies of electric motors compared to internal combustion heat engines. Fig. 7.12 also highlights the reality that any system that includes the use of a heat engine will have relatively low total efficiencies. However, the generation of electric power from coal, as seen in earlier chapters, is a relatively inefficient step, so when it is considered, the system efficiency drops to the point of only being marginally superior to a gasoline-powered vehicle (and essentially even, if we considered one of the better diesel cars). If instead of coal, the EV is charged using electricity generated by a combined-cycle natural gas power plant, the increased thermal efficiency now translates to significantly higher mpg$_e$ value.

If the EV is now charged solely from using electricity generated from hydropower, the mpg$_e$ becomes much more significant. Hydropower is probably the

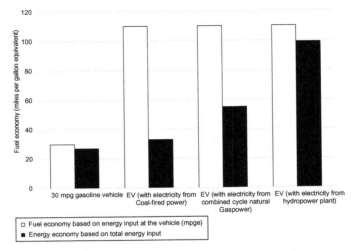

FIGURE 7.12

Mile per gallon equivalents for gasoline-powered ICE and electric vehicles, considering energy input at the vehicle and total energy input, based on the electric power system. Calculations assume 10%, 30%, 50%, and 90% efficiency from electric power system to engine or motor. *ICE*, internal combustion engine.

most efficient energy source that humanity employs: commonly converting 90% of the input energy to electricity in large-scale hydropower facilities. If it is the primary source converted to electricity, then EVs have the potential to be far more energy efficient in total than their ICE counterparts. This is an important consideration when looking at EVs (and PHEVs), because most utility grids around the world continue to increase in efficiency by relying less on the least efficient of the fossil fuels. At that point, the primary issue to widescale adoption becomes limitations in driving range due to low battery capacity, if we want to use EVs exactly like we do ICEs. The ICE model may not necessarily be one that we wish to continue. Indeed, since there is a very wide range in fuel economies for personal vehicles, the selection of a vehicle is one of the most important consumer actions in determining driving efficiency. Of course, the choices are a function of individual needs. Matching needs to the vehicle best suited to meet those needs is desirable.

7.4 DRIVING MORE EFFICIENTLY

The overall efficiency of vehicle operation is a function of the fuel used, the vehicle and its appropriateness for the transportation-specific function, vehicle accessories and convenience features, road quality, and operator driving behavior.

As described in the joint US DOE/EPA's fuel economy information website, fueleconomy.gov, operator driving behavior is one of the most significant controllable factors in vehicular transportation on roads.[3] The main factors listed include:

- uniform driving patterns,
- moderating speed,
- eliminating cargo on the vehicle roof,
- maintenance (i.e., tuning the engine, maintaining tire pressure, using appropriate motor oil), and
- minimizing excess weight.

DOE/EPA refers to uniform driving patterns as driving "sensibly," which can also be described in the negative as not driving "aggressively" or erratically. The most common aspect of this is the tendency to accelerate rapidly from a stop—and then often brake sharply at the next stop sign or traffic light. It is clearly wasteful to brake sharply, as the braking process, by design, employs friction to slow the vehicle. This converts the energy of the rolling vehicle into heat to be dissipated. There can be no doubt that the most efficient driving pattern would be to maintain a constant speed, without ever braking until the destination is reached. While this is generally not possible, it does represent the ideal driving pattern to approach. Studies by DOE/EPA suggest that "aggressive" driving is likely to lower fuel economy by 15%–30% in highway conditions and 10% to as much as 40% in stop-and-go driving conditions. It can be accomplished first by avoiding rapid acceleration and then by being alert to the road and what lies ahead. While automobiles must be capable of reasonably vigorous acceleration to facilitate merging with high-speed traffic on highways, this capability makes it very easy to over-accelerate in routine stop-and-go driving. Rapid acceleration increases the power consumption (again, energy consumed per unit time) and reduces the efficiency of the engine. This is, of course, an even much larger factor if it leads to wasting most of that energy in sudden braking at the next regulated intersection.

Moderating speed is the next controllable operator behavior. Speed, itself, incurs an energy cost, related to increased air resistance. US DOE/EPA data suggest that there is an efficiency loss of approximately 10% for each additional 5 mph above 50 mph (8 kph above 80 kph). Below 50 mph, the effect is much smaller. The effect is primarily a result of increased drag associated with moving more rapidly through the air—the car is effectively driving into a wind created by the car's velocity. The use of cruise control facilitates maintaining a moderate speed. It produces the additional related benefit of more uniform driving (with less acceleration and deceleration).

Avoiding unnecessary idling is another important consideration, because, by definition, when a vehicle is idling, it is consuming energy, with no useful

---

[3]A significant and growing amount of energy is consumed in private automobiles, so much of the following discussion is focused on passenger vehicles, but a number of factors are applicable to long-haul trucking as well.

(transportation) work being performed. Indeed, one of the measures adopted by some automakers to enhance efficiency performance is to turn the engine off when idling and then automatically restart when the driver steps on the accelerator: often called "start—stop technology." This automated approach to reducing idling was first introduced in Europe and recently has become available in the United States. Without this technology, it is recommended to turn off one's engine when stopped for more than a few seconds: the US DOE/EPA notes from a study by Argonne National Labs that it can save fuel even if for as little as 10 seconds. Of course, idling a car for prolonged periods can waste a considerable amount of fuel. For drivers of private cars, this is largely discretionary, but it is less so for long-haul trucks. Heating or cooling the cab when pulled off for the driver to sleep is the source of a great deal of idling fuel consumption.

With respect to avoiding cargo on the vehicle roof, roof racks are a popular feature added to many vehicles, but, being outside of the frame of the car, anything loaded on the racks adds to the surface area moving through the air, thus increases air resistance. Unless a specifically designed cargo bin is used, the loading is likely to have a poor aerodynamic profile. Anything that increases the surface profile of a vehicle increases its air resistance, which decreases its energy efficiency.

Proper maintenance is a key factor to ensure that any vehicle is actually performing to its specifications. It cannot achieve its expected performance if the engine components are not working together, as designed. "Tuning" an engine has taken on different features as cars moved from carburetion to electronic fuel injection. In either case, the engine must be operating properly to achieve efficient energy conversions. The process is relatively complex: fuel and air are delivered to the cylinder, which compresses it to the top of the compression stroke, where a spark plug ignites the fuel/air mixture. All of this must be properly controlled to ensure efficient combustion and conversion of the heat to mechanical power. Improper air/fuel mixtures or quantities and improper timing can adversely affect performance. Maintaining appropriate tire pressure also falls into this category, since tires provide for traction to permit independent maneuvering. The softer they are, the more contact there is with the pavement, which contributes to greater rolling friction losses. Lastly, using appropriate motor oil is important for maintenance. Motor oil serves the role of reducing friction between the moving parts in the engine. Friction, necessarily, converts kinetic energy to heat, which is wasted. The motor oil must also, though, perform appropriately through the reasonable range of engine operating conditions. It must provide sufficient viscosity to lubricate the engine's moving parts at high operating temperatures, while not being too thick to flow effectively at low start-up conditions.

Finally, related to this is minimizing excess weight, since weight is primarily an effect on fuel economy for accelerating the vehicle. This can include, though, not just increasing speed but hill climbing as well.

Fig. 7.13 illustrates some typical ranges of likely savings, based on the several efficient driving practices just described. In each category, the black bar

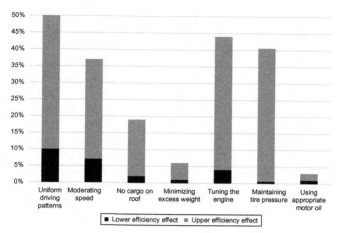

FIGURE 7.13

Estimated range of energy-saving potential for several driving behavior options based on a DOE/EPA analysis.

Figure created using data presented in US DOE/EPA's fueleconomy.gov, Driving More Efficiently (2018).

represents modest changes, while the gray bar represents a high estimate of what an extreme change in behavior might constitute. The ranges depend on the extremes assumed. For instance, the relatively small bar for "minimizing excess weight" is based on a range of 100–500 lb of excess weight in the vehicle. For the seven choices in vehicle operations depicted in Fig. 7.13, it is clear that the largest factor is literally how one drives: to maintain relatively uniform speed, with minimal aggressive acceleration. These combined effects can account for nearly a 50% improvement in efficiency or fuel economy, relative to extremely aggressive driving patterns. It is important to note that there is overlap in several of the behaviors shown, as, for instance, maintaining a moderate speed would be highly correlated with maintaining uniform driving patterns. It is clear that keeping the vehicle's engine tuned and tires properly inflated will generally have little effect; however, they can have a very large efficiency cost if seriously ignored.

Other considerations with respect to driving efficiency include planning and combining trips and strategies for dealing with driving in cold weather. With respect to the former, it is quite apparent that combining trips between people (e.g., in carpooling) saves energy. The efficiency can be considered to improve in that the practically useful work (number of people being moved), increases. Similarly, planning to stop at the grocery store on the way home from work, rather than making an extra trip, can save energy. The performance of any vehicle can be affected by weather, especially extremes in temperature. The fuel and the motor oil are more viscous in low temperatures, which can require more energy to pump them and, in the case of motor oil, can mean that the oil is not circulating to coat and lubricate the engine (especially the piston cylinder walls) effectively in the time before the engine warms up. An enclosure, such as a garage (particularly heated or attached to a heated home), will keep the vehicle from getting very cold.

Idling is particularly an issue in cold weather, as there is a strong temptation for drivers to idle their cars to warm up. Part of this is a comfort issue and is a significant motivation to purchase "remote starters" so that the car can be comfortably warm when the driver (and passengers) enters it. So, the task in this fuel consumption changes from transportation to space heating of the vehicle in this case. In addition, since a cold engine runs less efficiently than a warm one, drivers can, with some justification, think that idling the car to warm up the engine is good for efficiency. In fact, though, the engine warms quickly in use and only a brief warm-up is desirable before driving off (albeit not aggressively).

Seat warmers, and even heated steering wheels now, offer another solution to providing comfort relatively quickly in cold weather. Of course, the use of all accessories draw on the vehicle's electric system, which can increase the load on the alternator in an ICE vehicle or directly compete with driving the vehicle for the use of battery power in an electric or hybrid vehicle. In hot weather, on the other hand, cooling the vehicle is the challenge to maintain comfort. The air conditioner operated by an ICE is particularly inefficient.

As mentioned previously, proper tire pressure is a factor in driving efficiency and can require more frequent checking in cold weather.

7.4.1 TIPS FOR HYBRID GASOLINE/ELECTRIC VEHICLES AND ELECTRIC VEHICLES

According to a paper by Thomas, EVs and various forms of gasoline—electric hybrids have some special considerations, in addition to those that are true for all vehicles. Just as in the case of ICE vehicles, sharp braking dissipates a large amount of kinetic energy as waste heat. The additional consideration for electric and hybrid vehicles is that they may have regenerative braking capabilities that effectively use the internal resistance of a dynamo to slow the car in gradual stopping conditions, converting much of the kinetic energy back into electricity to be stored in the battery. Since it is sometimes necessary to stop a vehicle quickly in a potential emergency, it is necessary to have conventional friction brakes as a backup. Sudden stops will engage the conventional brakes, negating the potential energy captured from regenerative braking.

The electric motors in HEVs, including plug-in HEVs, and EVs operate much more efficiently than the ICE, so it is advantageous to operate in the electric mode as much as possible. This means that the battery should be sufficiently charged at the start of every trip so that the ICE does not need to engage at all in short trips. The energy contained in a relatively small quantity of gasoline or diesel can greatly enhance the cruising range, which largely accounts for the popularity of hybrids, but the most efficient operation of the vehicle entails only tapping that ICE capacity when long cruising range is needed (and the ICE can operate at its most effective in maintaining speed for cruising.)

Finally, the electronic accessories all tap into the battery, which we have seen does not provide particularly dense energy storage. The more that they are used, the less driving power is available from the battery.

7.5 CALCULATION: FUEL ECONOMY AND PER PASSENGER EFFICIENCY

As we have explained throughout this chapter, it is helpful to contextualize fuel economy and/or energy intensity across multiple modes. It is also helpful to do this on a per person or per cargo basis, because it allows one to compare the efficiency of different modes with respect to how "full" a vehicle is, because a substantial amount of wasted fuel occurs when a vehicle is not at full load or capacity to accomplish its intended task: transport of people or cargo. This may seem less important to individuals driving in private automobiles that can afford driving at low capacity; however, large operations, such as airlines or trucking and busing companies, have a financial incentive to operate at full capacity. Thus devising a methodology to calculate per-person or per-cargo fuel economy provides an informative metric with respect to energy efficient transportation policy and planning.

The ORNL has devised such a methodology to estimate average per-passenger fuel economy in miles per gallon of gasoline equivalent (GGE) for the United States as a whole, using the following variables:

- Number of vehicles for a given year
- Vehicle-miles traveled in that year
- Load factor, defined as an average number of persons per vehicle
- Passenger-miles, defined as vehicle-miles multiplied by the load factor
- Energy used, in units of GGE, using the heat content of individual fuels and electricity to compare gasoline, diesel, and electricity on an even basis

To calculate the per-passenger fuel economy, the ORNL simply divides passenger-miles by GGE, to arrive at the metric *passenger-miles per gallon of gasoline equivalent* (pmpGGE). Fig. 7.14 summarizes the ORNL analysis of load factors and pmpGGE for common domestic modes.

Fig. 7.14 demonstrates several interesting features with respect to US passenger transportation modes. Rail is the most passenger-efficient transport mode due to relatively high load factors and relatively efficient rail transport. Domestic airlines have become extremely passenger transport efficient, especially in the last 10–15 years, which ORNL attributes to fitting more passengers onto planes and aggressively filling planes to capacity. Motorcycles have low load factors but have high fuel efficiency, making them very passenger efficient. Transit buses,

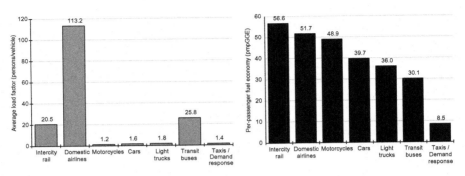

FIGURE 7.14

Average 2017 load factors and pmpGGE for several US transport modes. Combining load factor (passengers per vehicle) and equivalent fuel economy also for a more accurate comparison of the efficiency of the domestic transport modes. *pmpGGE*, Passenger-miles per gallon of gasoline equivalent.

Figure created with source data presented in ORNL's Transportation Energy Databook, Edition 37 (2018).

and taxis, on the other hand, have some of the lowest passenger efficiencies. Taxis are inefficient by nature, due to a naturally low occupancy and having to use fuel to collect passengers and to return to a centralized location. The ORNL reports that transit buses only tend to be 25% full on average in the United States, making them less efficient than driving a car or light truck, thus illustrating that ridership is very important. Finally, based on Fig. 7.14, cars and light trucks actually have a moderate passenger efficiency and, given their scheduling flexibility (passengers are not "constrained" by bus, rail, or airline routes and schedules), there is an obvious cost-benefit to their use. Note that by adding just one additional passenger onto cars, the pmpGGE would roughly increase by almost 60%, potentially making it competitive with intercity rail at its typical load factors.

One can take this a step further to include other modes, including cargo (by estimating equivalent passenger to cargo weight) and even for modes that do not require commercial fuels, such as bicycles or even walking (by estimating average speeds and the caloric intake needed to move one's body). Fig. 7.15 shows passenger and cargo transport efficiency for several additional modes that we calculated using a methodology similar to ORNL, considering average load factors (black) as well as maximum possible load (gray).

Fig. 7.15 confirms that from an equivalent passenger perspective, both heavy truck and freight train (i.e., rail) are extremely efficient modes of transport, especially when loaded to capacity. For regular passenger transport, substantial efficiency gains can be made by incentivizing ride-sharing in automobiles, especially in PHEVs. Finally, though not the focus of this chapter, walking and bicycling both offer very efficient and effective modes of transport, especially over short distances. Making roads and lanes available for walking and biking can have a significant positive impact on reducing fuel consumption worldwide.

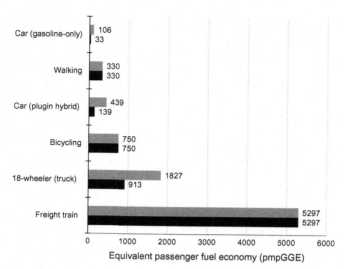

FIGURE 7.15

Equivalent passenger fuel economy for various transport modes. Freight transport modes were converted to passenger equivalent values by assuming one US ton was equal to 12.5 passengers. Walking and bicycling were estimated using average caloric intake and speed.

Figure partially based on data presented in US BTS's Energy Intensity of Passenger Modes (2017), Average Fuel Efficiency of US light duty vehicles (2017), and ORNL's Transportation Energy Databook, Edition 37 (2018).

REFERENCES AND FURTHER READING

Books and Technical Articles/Reports

Aubrecht, G. J. (2006). *Energy: Physical, environmental, and social impact (3rd ed.).* Pearson Prentice Hall.

Ebenhack, B. W., & Martínez, D. M. (2013). *The path to more sustainable energy systems: How do we get there from here?* Momentum Press.

Fay, J. A., & Golomb, D. S. (2012). *Energy and the environment.* Oxford University Press.

Hinrichs, R., & Kleinbach, M. (2012). *Energy: Its use and the environment.* Nelson Education.

Irimescu, A., Mihon, L., & Pădure, G. (2011). Automotive transmission efficiency measurement using a chassis dynamometer. *International Journal of Automotive Technology, 12*(4), 555–559.

Masters, G. M., & Ela, W. P. (2008). *Introduction to environmental engineering and science.* Pearson.

Romer, R. H. (1976). *Energy: An introduction to physics.* Freeman Press.

Sivak, M., & Schoettle, B. (2012). Eco-driving: Strategic, tactical, and operational decisions of the driver that influence vehicle fuel economy. *Transport Policy, 22,* 96–99.

Thomas, J., Huff, S., West, B., & Chambon, P. (2017). Fuel consumption sensitivity of conventional and hybrid electric light-duty gasoline vehicles to driving style.

SAE International Journal of Fuels and Lubrication, *10*(3), 1—18. Available online at https://afdc.energy.gov/files/u/publication/fuel_consumption_sensitivity_style.pdf.
Energy Information Sources and Technical Reports
Fenske, G. (2009). *Parasitic energy losses.* Argonne National Laboratory. Available online at https://www.energy.gov/sites/prod/files/2014/03/f13/vss_17_fenske.pdf.
International Energy Agency. *Energy efficiency indicators: Fundamentals on statistics.* (2014). Available online at <https://webstore.iea.org/energy-efficiency-indicators-fundamentals-on-statistics>.
International Energy Agency. *Atlas of energy: Energy balance flows.* (2016). Available online at <https://www.iea.org/Sankey/>.
International Energy Agency. *The IEA energy efficiency indicators database.* (2017a). Available online at <https://www.iea.org/newsroom/news/2017/december/the-iea-energy-efficiency-indicators-database.html>.
International Energy Agency. *World energy outlook.* (2017b). Available online at <https://www.iea.org/weo/>.
International Energy Agency. *Energy efficiency indicators: Highlights.* (2018a). Available online at <https://webstore.iea.org/energy-efficiency-indicators-highlights>.
International Energy Agency. *Tracking clean energy progress: Transport.* (2018b). Available online at <https://www.iea.org/tcep/transport/>.
International Energy Agency. *Transport.* (2018c). Available online at <https://www.iea.org/topics/transport/>.
International Energy Agency. *The future of rail.* (2018d). Available online at <https://www.iea.org/futureofrail/>.
National Academies Press. *Review of the 21st century truck partnership.* (2008). Available online at <https://www.nap.edu/catalog/12258/review-of-the-21st-century-truck-partnership>.
Oak Ridge National Laboratory (ORNL). *Transportation energy data book, edition 37.* (2018). Available online at <https://cta.ornl.gov/data/index.shtml>.
Sperling, D., Lutsey, N., & National Academy of Engineering. (2009). Energy efficiency in passenger transportation. *The Bridge*, *39*(2). Available online at https://www.nae.edu/19582/Bridge/EnergyEfficiency14874/ThePotentialofEnergyEfficiencyAnOverview.aspx.
U.S. Energy Information Administration, International Energy Outlook. *Transportation energy consumption.* (2016). Available online at <https://www.eia.gov/outlooks/ieo/>.
US Department of Energy Office of Energy Efficiency and Renewable Energy. *Electric vehicles: Saving on fuel and vehicle costs.* (2018a). Available online at <https://www.energy.gov/eere/electricvehicles/saving-fuel-and-vehicle-costs>.
US Department of Energy Office of Energy Efficiency and Renewable Energy. *Parasitic loss reduction research and development.* (2018b). Available online at <https://www.energy.gov/eere/vehicles/parasitic-loss-reduction-research-and-development>.
US Department of Energy Office of Energy Efficiency and Renewable Energy, Alternative Fuels Data Center. *Alternative fuels and advanced vehicles.* (2018a). Available online at <https://afdc.energy.gov/fuels/>.
US Department of Energy Office of Energy Efficiency and Renewable Energy, Alternative Fuels Data Center. *How do gasoline cars work?* (2018b). Available online at <https://afdc.energy.gov/vehicles/how-do-gasoline-cars-work>.
US Department of Energy Office of Energy Efficiency and Renewable Energy, Alternative Fuels Data Center. *Vehicle weight classes & categories.* (2018c). Available online at <https://afdc.energy.gov/data/10380>.

US Department of Energy Office of Energy Efficiency and Renewable Energy and US Environmental Protection Agency. *Driving more efficiently*. (2018a). Available online at <https://www.fueleconomy.gov/feg/driveHabits.jsp>.

US Department of Energy Office of Energy Efficiency and Renewable Energy and US Environmental Protection Agency. *Where the energy goes: Gasoline vehicles*. (2018b). Available online at <https://www.fueleconomy.gov/feg/atv.shtml/>.

US Department of the Interior, Bureau of Reclamation. *Reclamation: Managing water in the west: Hydroelectric power*. (2005). Available online at <https://www.usbr.gov/power/>.

US Department of Transportation Bureau of Transportation Statistics. *Average fuel efficiency of US light duty vehicles*. (2017a). Available online at <https://www.bts.gov/content/average-fuel-efficiency-us-light-duty-vehicles>.

US Department of Transportation Bureau of Transportation Statistics. *Energy intensity of passenger modes*. (2017b). Available online at <https://www.bts.gov/content/energy-intensity-passenger-modes>.

US Energy Information Administration. *Today in energy: Fuel economy standards have affected vehicle efficiency*. (2012). Available online at <https://www.eia.gov/todayinenergy/detail.php?id = 7390>.

US Energy Information Administration. *Global transportation energy consumption: Examination of scenarios to 2040 using ITEDD*. (2017). Available online at <https://www.eia.gov/analysis/studies/transportation/scenarios/pdf/globaltransportation.pdf>.

US Energy Information Administration. *Energy explained: Energy use for transportation*. (2018a). Available online at <https://www.eia.gov/energyexplained/?page = us_energy_transportation#tab1>.

US Energy Information Administration. *Total energy: Monthly energy review*. (2018b). Available online at <https://www.eia.gov/totalenergy/data/monthly/>.

US Environmental Protection Agency and National Highway Traffic Safety Administration. *Greenhouse gas emissions standards and fuel efficiency standards for medium- and heavy-duty engines and vehicles*. (2011). Available online at <https://www.govinfo.gov/content/pkg/FR-2011-09-15/pdf/2011-20740.pdf>.

US Environmental Protection Agency. *Light-duty vehicle CO_2 and fuel economy trends*. (2018). Available online at <https://www.epa.gov/fuel-economy-trends/highlights-co2-and-fuel-economy-trends#Highlight1>.

World Bank. *Air transport and energy efficiency*. (2012). Available online at <http://siteresources.worldbank.org/INTAIRTRANSPORT/Resources/TP38.pdf>.

Other Resources

Greater New Haven Clean Cities Coalition. (2016). *What are the various vehicle weight classes and why do they matter?* Available online at <http://nhcleancities.org/2016/04/various-vehicle-weight-classes-matter/>.

Gross, D. (2015). Start-stop technology: It shuts down the car engine to save gas and it's coming to America. *Slate Magazine*. Available online at https://slate.com/business/2015/08/start-stop-technology-it-shuts-down-car-engines-to-save-gas-and-its-coming-to-america.html.

International Air Transport Association. *Traveler numbers reach new heights*. (2017). Available online at <https://www.iata.org/pressroom/pr/Pages/2018-09-06-01.aspx>.

International Council on Clean Transportation. *Fuel efficiency trends for new commercial jet aircraft: 1960 to 2014*. (2015). Available online at <https://www.theicct.org/sites/default/files/publications/ICCT_Aircraft-FE-Trends_20150902.pdf>.

International Council on Clean Transportation. *Light-duty vehicle greenhouse gas and fuel economy standards*. (2017). Available online at <https://www.theicct.org/sites/default/files/publications/2017-Global-LDV-Standards-Update_ICCT-Report_23062017_vF.pdf>.

International Organization of Motor Vehicle Manufacturers. *2017 Production statistics*. (2018). Available online at <http://www.oica.net/category/production-statistics/2017-statistics/>.

Morris, H. (2017). How many planes are there in the world right now? *The Telegraph*. Available online at https://www.telegraph.co.uk/travel/travel-truths/how-many-planes-are-there-in-the-world/.

National Research Council. (2015). *Cost, effectiveness, and deployment of fuel economy technologies for light-duty vehicles*. National Academies Press. Available online at https://www.nap.edu/catalog/21744/cost-effectiveness-and-deployment-of-fuel-economy-technologies-for-light-duty-vehicles.

North American Council for Freight Efficiency. *Current technology*. (2018). Available online at <https://nacfe.org/current-technology/>.

Plumer, B., & Popovich, N. (2018). How US fuel economy standards compare with the rest of the world's. *The New York Times*. Available online at https://www.nytimes.com/interactive/2018/04/03/climate/us-fuel-economy.html.

Sprinkle, T. (2018). *Improving the internal combustion engine, part 1*. American Society of Mechanical Engineers. Available online at https://www.asme.org/engineering-topics/articles/energy/improving-internal-combustion-part-1.

US National Aeronautics and Space Administration, Glenn Research Center. *How does a jet engine work?* (2014). Available online at <https://www.grc.nasa.gov/www/k-12/UEET/StudentSite/engines.html>.

Residential and commercial sector energy efficiency

8

CHAPTER OUTLINE

The residential and commercial sectors contain those elements of society with the most direct impact on the everyday lives of the world's population. In this chapter, we review:

- Major services and functions;
- Common devices;
- Energy processes and efficiencies within the most utilized component: buildings.

8.1 RESIDENTIAL AND COMMERCIAL SECTOR OVERVIEW

The residential and commercial sectors are key components of local and regional economies.[1] They are where people live and work, congregate for school and social activities, house governments, as well as seek care for medical services and religious functions. Buildings are perhaps the most obvious and tangible feature

---

[1]The International Energy Agency (IEA) uses the word "services" instead of "commercial" but represents the same or similar entities.

Energy Efficiency. DOI: https://doi.org/10.1016/B978-0-12-812111-5.00008-1

of the residential and commercial sectors. They are relied upon to provide shelter from the outside elements, and to provide energy-related needs, such as thermal comfort, illumination, communication, refrigeration, and even varied forms of entertainment. The desire to improve efficiency within these structures stems mostly from optimizing the energy conversions used to satisfy these tasks. Along with industry and transportation, the residential and commercial sectors are classified as end-use sectors, utilizing large amounts of fuels and electricity to service many basic human needs. As noted in Chapter 6, Industrial sector energy efficiency, the annual energy consumption from these two sectors is fast approaching 90 quads ($\sim$95 EJ).

Residential and commercial sectors are distinguished by the scope, scale, and purpose of the buildings within each sector. The US Energy Information Administration (EIA) defines the residential sector as "living quarters for private households," and the commercial sector as "service-providing facilities and equipment of businesses, federal, state, and local governments, and other private and public organizations, such as religious, social, or fraternal groups." Essentially: homes, apartments, businesses, restaurants, government buildings, hospitals, schools, and various private and public organizations, and public works facilities. Despite structural differences, it is reasonable to treat the residential and commercial sectors as one combined sector because they are intertwined in local communities and share common end uses of energy, including:

- space heating and cooling,
- water heating,
- lighting,
- refrigeration,
- cooking,
- appliances, and
- miscellaneous devices.

The combined residential and commercial sector accounts for a significant share of total and end-use energy consumption in OECD (Organisation for Economic Co-operation and Development) countries. The predominant fuels and energy sources used in these sectors are natural gas, oil, district heat, and electricity, which are used to satisfy the energy needs of local and regional communities, mostly related to local climatic conditions. (Coal and biomass are also used extensively in the developing nations.) Recent residential and commercial end-use energy consumption data for the United States can be used as an example of typical energy end-use distributions in the OECD. These data are regularly gathered by the EIA through their Residential Energy Consumption Survey (RECS) and their Commercial Building Energy Consumption Survey (CBECS). The United States is a good example because it spans a representative variety of climates from temperate to subtropical, arid to humid, as well as arctic in Alaska.

As shown in Fig. 8.1, the most recent 2015 US RECS found that space heating dominates US residential sector energy end use, using roughly 43% of the

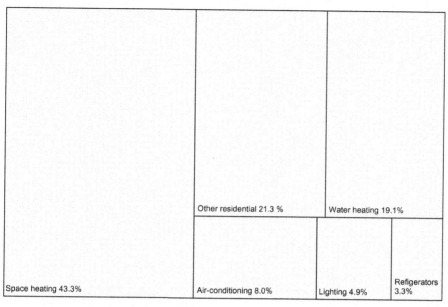

	Other residential 21.3 %	Water heating 19.1%
Space heating 43.3%	Air-conditioning 8.0%	Lighting 4.9%	Refrigerators 3.3%

FIGURE 8.1

Energy end uses in the US residential sector, 2015. Total US residential end-use energy consumed in 2015 was 11.3 quads (11.9 EJ). "Other residential" includes TVs and related, clothes dryers, ceiling fans, air handlers, cooking, dishwashers, clothes washers, humidifiers/dehumidifiers, and other smaller appliances and electronic devices.

Figure created using data presented in the US EIA 2015 Residential Energy Consumption Survey.

reported 11.3 quads of end-use energy in 2015. Air-conditioning (i.e., space cooling) used roughly 8%, and when added to the share from space heating, the combined space heating and cooling category represented more than half (51%) of the total 2015 energy consumption in the residential sector. In addition, the 2015 RECS found that water heating (used for bathing, washing, household cleaning, and laundry) consumed 19% of all end-use energy, followed by lighting (5%) and refrigerators (3.3%). "Other residential" (21.3%) includes TVs and related equipment, clothes dryers, ceiling fans, air handlers (for heating and cooling), cooking, dishwashers, clothes washers, humidifiers/dehumidifiers, other smaller appliances and electronic devices.

For comparison, Figs. 8.2 and 8.3 present recent data for OECD (of which the United States is a member) and Chinese residential end-use consumption, respectively. Though categorized slightly differently, the data show interesting similarities to the United States. In particular, most of the OECD and Chinese energy consumption also went to building heating and cooling, comprising 55% and 56% of total end use, respectively. This is similar to the US percentage (51%). Other end uses also line up well between OECD and the United States, whereas China deviates

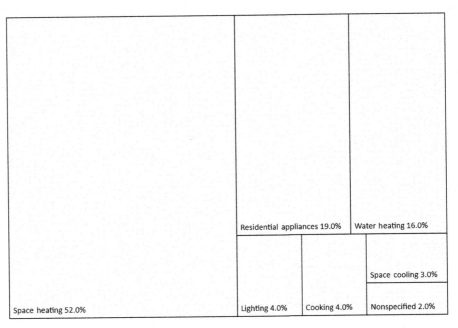

FIGURE 8.2

Energy end uses in the OECD residential sector, 2013. The breakdown of energy consumption is very similar to US data.

Figure created using data presented in IEA Energy Efficiency Indicators: Fundamentals on Statistics (2014).

significantly from both OECD and the United States in cooking. Being a developing nation, this should not be a major surprise, since there remains a significant rural population with minimal access to electricity and other energy-intensive services, causing much of the population to continue to use firewood to cook.[2] Another notable difference between the United States and much of the OECD is that the amount of end-use energy used in the United States is substantially higher. Whereas the United States consumes nearly 12 EJ of energy just for end use in the residential sector, many highly developed countries in the OECD consume between 0.1 and 1 EJ, as populations are much smaller in individual member nations.

With respect to commercial sector end-use energy consumption, the most recent 2012 US CBECS revealed notable end-use differences compared to

---

[2]Many developing nations are structurally different—they are located in the tropics to subtropics; thus space heating is a much smaller demand. Space cooling is more universally desirable, but not a survival need. Water heating is also desirable, but not essential and commonly unavailable to most poorer countries. Indeed, cooking is the central need and dominates both commercial and residential energy demand among the developing nations.

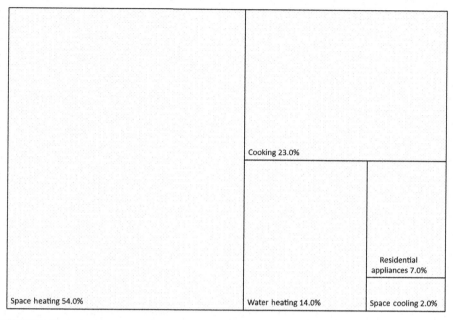

FIGURE 8.3

Energy end uses in China's residential sector, 2012. Space heating and water heating end-use consumption is very similar to OECD; however, a much higher percentage still goes to cooking. *OECD*, Organisation for Economic Co-operation and Development.

Figure created using data presented in Zheng et al. (2014).

US residential sector end use, depicted in Fig. 8.4. Whereas the residential sector used 51% combined for heating and cooling, the commercial sector used 44.2%—i.e., 25.2% for space heating and 19% for air-conditioning (a composite of cooling and ventilation). The difference in heating and cooling needs likely has to do with many commercial and government buildings being closed at night as well as the additional thermal loads that exist from solar gain and human body heat. Those require more cooling and ventilation. In addition, refrigeration and lighting are used more in the commercial sector, 9.6% and 10.4%, respectively. "Other commercial" (28.5%) includes cooking, which is a larger energy end user compared to the residential sector (likely for food and restaurant services), as well as computing, office equipment, medical equipment, and other devices.

In addition, since buildings are at the heart of the residential and commercial sectors, the key energy efficiency indicator tracked by both EIA and IEA is building energy intensity, defined as the amount of end-use energy consumed per unit area. This can be measured and/or modeled as total end use, or specific end use (space heating, electricity, etc.). Fig. 8.5 depicts US CBECS and RECS for total

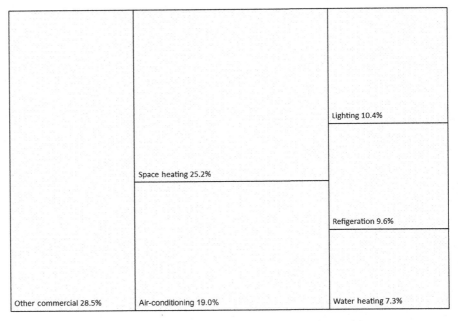

FIGURE 8.4

Energy end uses in the US commercial sector, 2012. Total US commercial end-use energy consumed in 2012 was 8.29 quads. "Other commercial" includes cooking, computing, office equipment, and other equipment and devices.

Figure created using data presented in the US EIA 2012 Commercial Building Energy Consumption Survey.

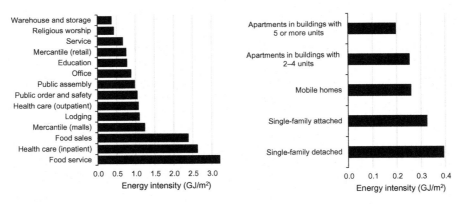

FIGURE 8.5

Total building end-use energy intensities in the US commercial and residential sectors in units of gigajoules per square meter (GJ/m^2). By far, the most energy-intensive buildings in the United States are related food service and sales, and inpatient health care.

Figure created using data presented in the US EIA 2012 Commercial Building Energy Consumption Survey and the US EIA 2015 Residential Energy Consumption Survey.

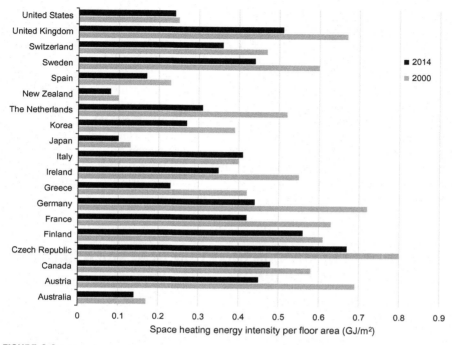

FIGURE 8.6

Years 2000 and 2014—space heating end-use energy intensities in the OECD residential sector in units of gigajoules per square meter (GJ/m²). *OECD*, Organisation for Economic Co-operation and Development.

Figure created with source data presented in IEA Energy Efficiency Indicators Database (2017).

building energy intensity in the years 2012 and 2015, respectively. Note that the most energy-intensive buildings in the commercial sector are food-related services and inpatient health care. This is most likely due to the need for continuous energy to heat and cool spaces, to power refrigerators and medical equipment, and other devices. On the residential side, the most energy-intensive buildings are single-family detached homes, which are about as energy-intensive as warehouses on the commercial side.

Since space heating is such an important component of end-use energy in the OECD, the International Energy Agency (IEA) tracks space heating intensity as its own indicator in its Energy Efficiency Indicator Database. Fig. 8.6 shows differences in residential space heating intensity from 2000 to 2014. Note that there is a wide span in space heating intensity in the OECD in general and that countries with some of the largest intensities have significantly reduced total end-use space heating intensity between 2000 and 2014. The IEA attributes this reduction to better thermal insulation of new buildings and thermal retrofit of older ones (to be discussed later).

Finally, it is notable that since the IEA has been tracking sectoral energy consumption in the OECD, while total final energy consumption has been rapidly rising around the world on a percentage basis, energy consumption in the residential and the commercial sectors has not changed since 1973. (IEA reports that the total final consumption in the combined residential and commercial sectors remains at 31% of the total.) As countries continue to transition to becoming service economies, with less emphasis on energy-intensive industry and more emphasis on transport and communication, it will be important to continue to track the combined sector's role in global consumption. New buildings continue to be erected globally, and to be able to maintain or even decrease energy consumption in these sectors, it will be important to continue to focus on energy efficiency in buildings, especially in the primary areas of space heating, water heating, and residential appliances.

8.2 END-USE ENERGY CONSUMPTION IN RESIDENTIAL AND COMMERCIAL BUILDINGS

Based on the combined sector overview presented in Section 8.1, the most important energy-intensive end uses in buildings in descending order are:

- space heating and cooling (including ventilation),
- water heating,
- appliances, including refrigeration,
- lighting,
- electronic devices, and
- cooking.

Of these end uses, the most building-integrated systems usually involve space heating and cooling, water heating, and lighting. These three combined comprise 75% of the US (and OECD) end-use residential energy consumption and 61% of the US end-use commercial energy consumption. Not only do these systems consume the most of the sector's end-use energy, they also are likely to remain in place for 20 or more years. Appliances, when aggregated, also make up a significant share of end-use energy, however, are less integrated but still necessary, and tend to remain in place 5–10 years. Given this analysis, a review of the most prominent end-use energy systems and processes are described in the following subsections.

8.2.1 SPACE HEATING AND COOLING

As stated earlier, the primary purpose of most dwellings is to protect humans from the outside elements. Air temperature, humidity, radiation, precipitation, and air movement—the five primary climatic variables—all have an immediate impact on human comfort. And, according to Krigger and Dorsi, humans feel comfortable when they are in a steady-state thermal balance with the local surroundings. In this steady state, the human body is losing as much heat as it is gaining from metabolic processes. Outside this state, humans will begin to feel discomfort and will tend to sweat or shiver. Air temperature is usually the primary climatic variable that determines human thermal comfort, with radiant temperature also a factor when indoors. And because humans spend so much time indoors, especially in the developed countries, we have increasingly come to expect our buildings to be equipped with heating and cooling systems necessary to *condition* spaces and maintain steady, comfortable temperatures.

Buildings also interact with the environment and, thus, will reach a steady thermal state with the outside air. Since buildings have a thermal mass, they will tend to absorb solar radiation, gaining thermal energy, and warming up. In addition to this solar gain, buildings also have internal gains from things inside of it, including people, pets, appliances, lights and other radiation sources. If the temperature outside is below an acceptable comfort range for the occupants, say in the winter, then it is necessary to heat the space from within using a space heater or multiple heaters until the inside temperature reaches a comfortable value. Alternatively, if the temperature outside is above an acceptable comfort range, say in the summer, then it is necessary to cool and/or ventilate the space using air-conditioning. The temperature difference between indoors and outdoors is the primary driving force for thermal transmission from within the inside space. Buildings also tend to have extra openings for ventilation (i.e., deliberate heat removal) and entry, such as windows and doors (also known as fenestration), as well as other leakage points through the building shell, which is often a function of the quality of building construction. Because of this, the building also is susceptible to transferring already conditioned air (heat or cooling) to the outside, which in turn requires extra heating or cooling of the space back to the desired temperature. A depiction of common heat inputs and losses for heating and cooling is provided in Fig. 8.7.

The ability of a building to perform the task of maintaining a set temperature using the least amount of fuel/electricity from the heater or cooler is a key concern for energy efficiency. This involves two separate but interdependent issues: heating and cooling system efficiency, and building design/performance. These will be addressed in the following subsections.

8.2.1.1 Space heating[3]

Space heating is defined by the IEA simply as the different means of heating spaces, which is accomplished by using a variety of fuels and heating systems.

---

[3]This section emphasizes residential space heating; however, many light-commercial operations use very similar heating systems. For larger commercial facilities, government building codes typically mandate forced ventilation for adequate worker health and safety. These buildings typically employ combined heating, ventilation, and air-conditioning (HVAC) systems as a result.

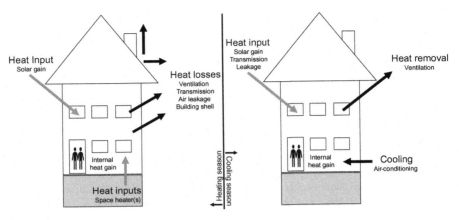

FIGURE 8.7

Heat inputs and losses in a building under space heating and cooling modes.

Figure created using composite information and depictions presented in Krigger and Dorsi (2009)
and Cullen and Allwood (2011).

The IEA separates heating systems into two broad categories: central heating and area-dedicated or room heating. Whereas dedicated room heaters heat just a room within a building, central heating systems are designed to heat the entire building. It is also quite common for occupants and/or households/businesses to use both central and area-dedicated systems, either during times of more moderate weather, or because the centralized systems are insufficient.

Central heating systems generate heat by transforming the chemical energy in fuel into thermal energy and transferring that energy to air, water, or steam, which then gets delivered throughout the building. The most common include:

- radiation systems (hot water or steam),
- hot air furnace systems (forced through walls or the floor via ducts and registers),
- district heating, and
- heat pumps.

The most common centralized heating systems in both residential and commercial spaces typically involve using a flame to heat and move a heat transfer fluid via forced hot air, pumped hot water, or steam delivery.[4] As per Krigger and Dorsi, these centralized oil, gas, and/or propane heating systems mix and burn fuel in a combustion chamber, and the heat from the flame and combustion gases are transferred via heat exchangers to the system heating fluid (air, water, or

---

[4]It is also possible to use electricity in a centralized furnace system, where fans move air heated by electric resistance heating elements through ducts. Though electric resistance heating is nominally 100% efficient, as much as 30% of that heat is lost through the distribution system.

steam). The combustion gases then leave the chamber through a flue, which typically connects to a chimney or other type of ventilation stack and the heating fluid is distributed throughout the space through pipes, ducts, vents, and/or metal baseboards and radiators, which is simply depicted in Fig. 8.8.

In addition, the overall efficiency of centralized heating systems is typically evaluated by using four criteria:

- Fuel burning efficiency—also known as the flame efficiency, or how much of a fuel's chemical energy is converted to heat at the flame.
- Steady-state efficiency—also known as the combustion efficiency, or how much heat gets transferred from the heat exchanger to the heat transfer fluids, accounting for fuel burning losses and chimney losses.
- Annual fuel utilization efficiency (AFUE)—how much fuel is actually utilizable and flows into the heating system ducts or pipes, accounting for fuel burning losses, chimney losses, stop/start cycling losses, and heat loss through the heating system cabinet.
- Seasonal efficiency—the delivered heating efficiency, accounting for distribution losses from the walls of ducts and pipes.

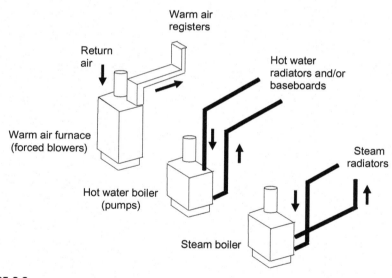

FIGURE 8.8

Simplified depictions of a warm air furnace, a hot water boiler, and a steam boiler centralized heating system. These systems are commonly found in colder climates. The heat transfer fluid (air, water, or steam) moves into the heating space via ducts/register and pipes/radiators/baseboards. The combustion gases leave via a flue and up a stack (chimney).

Figure based on a depiction presented in the American Council for an Energy-Efficient Economy's Smarter House online publication (2015), with modification.

Notice that seasonal efficiency is the most encompassing of the four criteria, which includes AFUE plus distribution losses; however, it is more difficult to measure. AFUE is the typical efficiency rating labeled on installed systems.

The US Department of Energy (DOE) *Energy Savers* Program highlights how to identify and compare a heating system's AFUE efficiency by installed components. According to the DOE, older, low-efficiency systems typically employ a natural draft that brings in combustion air in an imprecise manner, creating a constant flow of combustion gases that is easily lost out of the chimney. Low-efficiency systems also utilize a continuous pilot light, which is burning fuel even when not needed by the space, and the systems tend to be larger and heavier, especially the heat exchanger, causing extra cycling losses. All of these lead to 56%−70% AFUE. Midlevel efficiency heating systems improve AFUE by employing exhaust fan control to regulate the flow of combustion air and combustion gases more precisely. They also employ electronic ignition, with no pilot light, restricting fuel usage to only when the space needs to be heated. They are also more compact in size and of lighter weight, including smaller diameter flues, which helps reduce cycling losses. These additional improvements lead to 80%−83% AFUE. Finally, high-efficiency heating systems will additionally condense flue gases in a second heat exchanger to capture lost latent heat for extra efficiency. They will also have sealed combustion chambers, which combined leads to 90%−98.5% AFUE.

Much can be done to reduce AFUE losses via equipment choices; however, seasonal, or delivered heating efficiency (which accounts for distribution losses) can still be low. According to Krigger and Dorsi, because of the many ways that heat can be lost between the initial flame and the actual heat that makes it into the space, the delivered efficiency, when coupled to a low-efficiency heating system, can be as low as 35%. A table of loss estimates and processes is provided in Table 8.1.

Table 8.1 Combustion Heating System Efficiency Types, Potential Causes for Losses, and Percent Loss Ranges for Each Loss Type

Efficiency Type	Potential Losses	% Loss Range
Fuel burn efficiency	Unburned fuel	<1
Steady-state efficiency	Sensible heat	1−3
—	Latent heat	2−12
—	Excess air	5−30
Annual fuel utilization efficiency	Off-cycle/load matching	7−15
Seasonal efficiency	Distribution	5−35

Note that sensible heat and latent heat losses occur any time a phase change occurs during combustion and/or steam processes. Also note that excess air and distribution losses can be the largest types of system losses.
Table created using data presented in Krigger and Dorsi (2009).

Area-dedicated heating systems generate heat by transforming either the chemical energy in fuel or electricity into thermal energy (via standalone combustion or wire resistance heaters) and transferring that energy directly into a room or small floor space within the building. According to the IEA, the most common room heating systems include:

- propane or natural gas stoves,
- wood and/or wood pellet stoves,
- coal/charcoal stoves,
- kerosene heaters, and
- electrical resistance space heaters.

Room combustion heaters technically operate like centralized furnaces and boilers; in that they place a fuel into a combustion chamber and burn it to produce heat, which will be delivered to the room or dedicated area via radiation or forced convection using a fan or blower, usually with exhaust gases being expelled through a flue and/or chimney.[5] Likewise, electric heaters send electrical current through resistance wires and deliver the resulting heat via radiation or forced convection using a fan. These room heaters, however, do not distribute the heat beyond their dedicated area, thus, can theoretically be more efficient than centralized systems because they do not suffer distribution losses. Their design, including their small size, however, means that they are insufficient for heating entire buildings. Also, biomass stoves (wood and wood pellet) have the potential for user-related factors that can both greatly diminish efficiency and greatly increase particulate emissions, affecting air quality and human health. (Open fireplaces, one of the oldest heating methods, are extremely inefficient and impractical to heat most buildings, especially in cold climates.)

8.2.1.2 Space cooling

The IEA defines space cooling as "all equipment used for cooling a living area." Like space heating, cooling is divided into two main categories: central cooling systems, and room-dedicated systems. Central air-conditioning systems operate very similarly to forced-air furnace systems, in which cooled air is fed into ducts with vent registers placed throughout the entire building. More commonly, wall or window air conditioners (ACs) are used to cool a dedicated space, usually just a single room. Other possible cooling systems include evaporative coolers, heat pumps operating in reverse "cooling" mode, or district cooling. Most of these systems require electricity for proper operation.

ACs operate under the same principle as refrigerators, using electricity to transfer heat from inside to outside by evaporating and condensing a refrigerant.

---

[5]Older designs of propane, natural gas, and some kerosene heaters allow for no venting from the living floor space. This can result in safety and health hazards.

According to the DOE *Energy Savers* program, the primary components of ACs are as follows:

- Evaporators—coils that remove heat and humidity from the air using a refrigerant. The evaporation of the liquid refrigerant "pulls" heat contained in the inside air, resulting in a reduction in room/building air temperature.
- Condensers—coils where the previously vaporized refrigerant condenses, releasing heat to the outside air.
- Compressors—pumps used to move the refrigerant from the evaporator, located inside, to the condenser, located outside.
- Blowers—fans that circulate air over the evaporator to move the chilled air into the space to be cooled.
- Fans—used to dissipate the heat away from the condenser.
- Filters—used to remove dirt, dust, and other particles from the air, which affects refrigeration cycle efficiency.
- Thermostats—inside controllers used to regulate cooling inside the building.

A schematic of a residential central air-conditioning system is depicted in Fig. 8.9.

In addition, there are four main types of ACs: (1) the afore-described central air-conditioning, which circulates air via ducts and vents (these are often "split" systems that can be part of combustion and electric furnaces); (2) room ACs, which are "packaged" systems (evaporator, condenser, and compressor in one boxed unit) and typically installed in windows to cool small areas; (3) ductless, "mini-split" ACs (condenser and compressor are packaged outside and a wall-mounted indoor blower unit connected to a furnace or heat pump); and (4) evaporative coolers, which cool outside air using evaporated water and circulates it through the house (particularly effective in arid climates).

AC efficiency is determined by the ratio of heat (in units of Btu or J) the system can remove per hour for each watt of electrical power it draws when plugged into and running from the building electricity supply. For room ACs this ratio is known as the energy-efficiency ratio (EER) and for central air-conditioning systems, the ratio is called the seasonal EER (SEER), sharing the same basic equation for calculation shown by the following equation:

$$\text{EER or SEER} = \frac{\text{Heat removed (Btu or J) per hour}}{\text{Electrical power drawn (W)}} \tag{8.1}$$

Since 2014, the DOE also rates room ACs with what they call a combined EER (CEER), which also includes power used when the unit is in standby mode (not running but still drawing some power until called on). According to the US Environmental Protection Agency *Energy Star* Program's Room AC Database, currently available room AC models have CEERs that range from 9.9 (with a 28,000 Btu/h cooling capacity) to 14.7 (between 14,000 and 18,000 Btu/h cooling capacity). The most common cooling capacity on the market is 6000 Btu/h with CEERs of about 12.1. All of these models are fairly energy efficient compared to

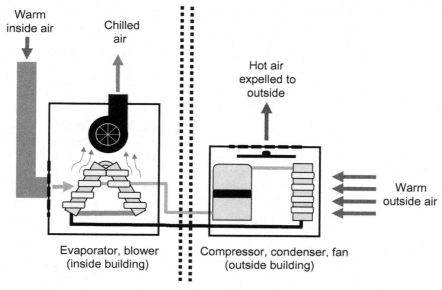

Warm
inside air

Chilled
air

Hot air
expelled to
outside

Warm
outside air

Evaporator, blower
(inside building)

Compressor, condenser, fan
(outside building)

FIGURE 8.9

Depiction of a centralized residential air-conditioning system. The evaporator and blower are located inside the building space, where warm air passes over the refrigerant coils, supplying heat for evaporation and cooling the air. A blower moves that chilled air into the room, and the refrigerant is pumped out to the compressor. Supply and return ducts within the house connect to a central air handler, typically located within a furnace housing. A fan is used to move the hot expelled air to the outside environment.

Figure created using information and imagery presented in the DOE Energy Savers Home Cooling Infographic (2016).

the less stringent standards of previous decades. (Note that an AC unit with a CEER of 12.0 typically requires half as much electricity to cool as an AC unit with a CEER of 6.0.)

The SEER is slightly different than EER because it is an average value that accounts for seasonal variations. This average value is estimated by setting the indoor temperature constant and measuring efficiency over a range of temperatures, typically between 16°C and 38°C. SEER values for central AC units on the market range from a minimum 13−14 to a maximum of 21−25. According to Krigger and Dorsi, pre-1979 SEER values of central AC systems available on the market ranged from 4.5 to 8.0. They note that the increases in energy efficiency compared to previous generation AC systems is the result of improved design, such as variable speed blowers and compressors, increased coil surface area, increase heat transfer to the air via better fin design, improved electric motor design, and evaporator fan control.

It is important to note that the SEER and EER on AC units represent a maximum rating. It is very likely that AC systems, in practice, will not achieve these maximum ratings, due to a number of factors, such as large variations in outside temperature between daytime and nighttime, user-related inefficiency by changing set point temperatures inside the building, and, perhaps most importantly, lack of proper maintenance and upkeep of the system. According to the US DOE, some of the most important maintenance tasks are to

- Replace and/or clean filters—Filters are used to keep dirt and dust out of AC system components, however, they need to be changed or cleaned routinely because clogged filters will obstruct airflow.
- Maintain clean evaporator and condenser coils—Even with filters, evaporator coils will collect dust and dirt over time. These reduce airflow and insulate the coils, which reduce cooling efficiency. Likewise, condenser coils, located outside, will tend to collect dirt, dust, and yard debris, all of which also affect efficiency.
- Repair damages coil fins—Aluminum fins located on the coils are used to more readily exchange heat between the air and the AC unit. These are easily bent and damaged, which reduces airflow to the coils. It is important to repair any coils that become damaged.
- Keep condensate drains clear—Clogged drains can lead to increased humidity, which is also an important factor in cooling.

Another important aspect of space cooling is ventilation. As mentioned previously, and as seen in Fig. 8.7, there are a number of thermal gains that occur within the building shell, primarily due to solar insolation in the summer, as well as internal heat gains from body heat and human activity, such as cooking, and heat produced from home appliances. In commercial buildings, additional heat gains may come from heat generated by machines and other larger devices. This heat (and humidity) buildup within the building can be removed from within the building shell via AC systems; however, it is often much more energy efficient to remove a large portion of this heat via ventilation. This can be done by taking advantage of a building's natural "stack effect," in which windows are opened at the top of a building to release the less dense hot air that has accumulated in that area. This is most effective in geographic areas that have large changes in temperature throughout the day. It can also be done mechanically via the use of fans, either placed in different parts of the buildings or in the roof/attic area, which help draw out the heated air. Fans consume much less electricity than AC systems and are quite effective in regions that are not always hot and humid; however, the DOE notes it is important to remember to turn area fans off when humans are not present, since fans do not cool spaces, only provide comfort to occupants. It is also important to reduce the impact of external and internal thermal gains, when relying on ventilation to help keep a building cool. Finally, in commercial buildings, it is often necessary to use ventilation in order to meet government mandates for continual input of minimum fresh outside air for air quality purposes. HVAC systems are an integral component to satisfy this requirement. Indeed, forced-air furnace heating systems are popular in both residential and commercial applications, because they allow for heating, cooling, and ventilation applications in a combined system.

8.2.1.3 Building thermal performance

As was explained earlier, achieving a desired temperature within buildings requires a thermal balance between external thermal gains and losses. Consider the cold weather scenario in Fig. 8.10. The main thermal gains are solar heat, internal heat (body heat, cooking, hot water usage, animals, etc.), and heat supplied by the heating system. The losses are primarily from air leaks and ventilation, wall and fenestration transmission (via conduction, radiation, and convection through walls and windows), and ceiling transmission (primarily via radiation).

Now consider the warm weather scenario in Fig. 8.11. There are gains due to the same factors (solar heat, air leakage, internal heat, and transmission) and purposeful losses primarily from the cooling system and from ventilation. Because

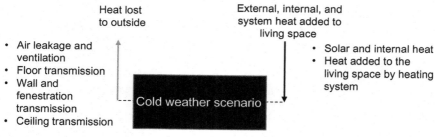

FIGURE 8.10

Depiction of building thermal gains and losses in a cold weather (heating) scenario.

Figure created using information presented in Krigger and Dorsi (2009).

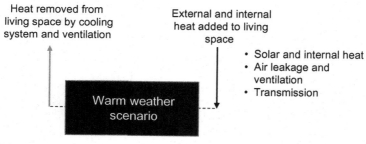

FIGURE 8.11

Depiction of building thermal gains and losses in a warm weather (cooling) scenario.

Figure created using information presented in Krigger and Dorsi (2009).

energy is required to add or remove heat from a living/working space, and because that energy typically comes in the form of fuel or electricity, reducing the energy needed to perform heating and cooling has a direct impact on energy costs and pollution.

Reducing building thermal energy needs is directly tied to a building's thermal performance, that is, how well the building can heat and cool using the minimum amount of fuel and electricity. Looking at Figs. 8.10 and 8.11, the primary factors that seem to most affect building performance are thermal transmission through ceilings, walls, and windows, and through conditioned air losses via air leakage. In order to reduce these losses, energy-efficient buildings will deliberately set a *thermal boundary*, via insulating materials and air-resistant sheeting materials (drywall, sheathing, siding, etc.). A tightly sealed building prevents air from crossing the boundary of the envelope (carrying heat or lack of heat with it), which reduces the convective heat loss or gain to the building. In addition, a well-insulated building will reduce conductive, radiative, and convective transmission losses through walls and act as a thermal resistance to temperature differences inside and outside of the building. In a new construction, it is easier to define the thermal boundary and use adequate materials to reduce heat losses through the building shell. In existing buildings, where it is desired to improve thermal performance, defining the boundary can be more challenging.

Fig. 8.12 depicts two choices for the thermal boundary of the structure originally depicted in Fig. 8.7. In residential buildings located in cold climates, the attic area can be one of the primary sources of air leakage and heat loss. It is often advantageous to define the thermal barrier at the ceiling located directly

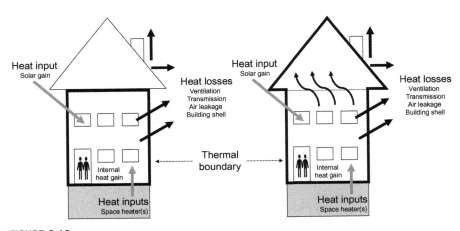

FIGURE 8.12

Depiction of the thermal boundary within a built structure. Insulation and air barrier applications for reducing heat transfer losses are impacted by boundary choice. It also provides better ideas for where breaks in the boundary may exist.

below the attic space. By doing this, one can minimize "thermal breaks" such as at eaves and at protrusions, such as chimneys. It may also be advantageous to include the basement as part of the thermal boundary in order to take advantage of the heat lost by the central heating system and allowing it to move up and into the living space. For homes that are located in warm, hot and humid, or hot and arid climates, it is also advantageous to reduce solar thermal gains by blocking solar radiation using light colored and/or reflective exteriors, and also by employing shading via landscaping (i.e., from shade trees) and/or window shading.

8.2.1.4 Recent trends in heating and cooling: heat pumps and insulation

Recall Fig. 8.7, which displayed space heating end-use energy intensity differences in the OECD residential sector from 2000 to 2014. There was a significant decrease in intensity in several countries, in particular, Germany, Ireland, Austria, and the Netherlands. The IEA attributes this reduction primarily to the broader adoption of two technologies in new builds and in building retrofits: electric heat pumps and insulation.

According to Krigger and Dorsi, whole-building heat pumps are essentially two furnaces packaged into a single unit. One furnace is a compressor that moves heat from the outside air to inside a building during the cold weather heating season using a refrigerant. The second furnace is an electric resistance heater, exactly like those found in electric furnace heating systems, used for backup or supplemental heat. Because heat pumps do not burn a fuel, but rather move heat from the outside and into a home, they can deliver up to four times more heat energy than the electrical energy they consume during operation (known as a coefficient of performance).

The most common type of heat pump system is an air-source heat pump, which uses two heat transfer coils to move heat from the outdoors into the indoors. Even in relatively cold temperatures, air has a certain heat content, so if a refrigerant being pushed outside through the coils is colder than the outside air, some of the outside air heat will be transferred to the coils, removing that heat. This relatively warmer air can then be pushed indoors via a fan or fans to help warm the inside space. This operates very similarly to an air-conditioning cycle, with the extra ability to allow refrigerant to heat or cool. Heating and cooling modes are depicted simply in Fig. 8.13. For many years, heat pump applications were limited to moderate climates; however, the US DOE reports that recent advances have allowed heat pumps to be effectively used in cold climates as well.

An increasingly popular variant of heat pump systems is the ductless, minisplit heat pump. According to the US DOE, ductless applications require minimal construction: a small hole is cut through a wall to connect the outdoor condenser and the indoor blower heads. These units are much smaller and tend to be more efficient since they have no ducts, reducing the potential for leaks and energy losses. They also tend to have better refrigerant charging. They are also particularly suited to open floor plans. A simple depiction is provided in Fig. 8.14.

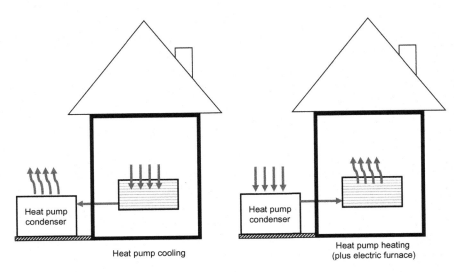

Heat pump cooling

Heat pump heating
(plus electric furnace)

FIGURE 8.13

Depiction of heating and cooling modes for a space heating/cooling heat pump system. During cooling mode, the heat pump transfers indoor heat outside, similar to an air-conditioning system. During heating mode, the system is reversed, extracting heat from the outside air and drawing it into the building. Air-sourced heat pumps typically also have electric heating elements to supplement the heating mode, especially in very cold weather. Conditioned air is transferred via fans and ducts.

Insulation is another key aid in improving a building's thermal performance and reducing building energy intensity. According to the Thermal Insulation Association of Canada, by definition, insulation performs the following thermal functions in residential and commercial applications:

- Conserves energy by reducing heat loss and gain (transmission).
- Reduces the prevalence of cold surfaces, thus preventing unwanted condensation.
- Increases the operating efficiency of heating, cooling, ventilation, plumbing, steam, process, and power systems.
- Controls surface temperatures for physical protection and comfort.
- Facilitates temperature control of system processes.
- Reduces or prevents exposure to fire or corrosive atmospheres.
- Reduces emissions of atmospheric pollutants by reducing fuel needs.

Focusing on residential buildings, insulation's primary purpose is tied to slowing heat transmission (conduction, radiation, and convection) through a building's walls, floor, ceiling, and roof. Of course, the driving force for heat transfer in these building is a combination of the temperature gradient between the inside space and the outside environment, as well as the thermal makeup of the building

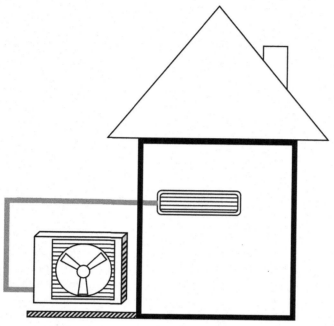

FIGURE 8.14

Depiction of a ductless, mini-split heat pump system. A relatively smaller compressor unit is located outside of the building and an indoor blower head (or multiple blower heads) are connected inside to move warm air into or out of the building.

materials. The more thermally insulating a material is, the slower heat transmission will occur, ultimately reducing heating (and cooling) requirements. Insulation is also key in reducing distribution losses found in building heating, cooling, and ventilation systems, both for indoor ambient air and for domestic water heating.

The most effective building insulation materials are those that are best at resisting heat transfer through them. The measure of a material's resistance to heat transfer is commonly known as the material's R-value. Specifically, R-value is physically represented by the ratio of a material's thickness divided by its thermal conductivity. Thus the higher the R-value, the better its insulating properties. R-value is an additive measure, meaning if you have two materials connected to each other, the R-value will be a composite of the R-values of each material.

Solid materials, such as steel, brick, and (to a lesser extent) wood, tend to be good conductors of heat, thus having low R-values. Gases, on the other hand, tend to be poor conductors of heat, thus having higher R-values. This is why most insulating materials are designed to maximize the amount of gas within them, either by using many fibrous strands that form millions of pockets of

Table 8.2 *R*-Values of Common Building Insulation Types in Units of *R*/in. and *R*/cm

Insulation Type	R/in.	R/cm	Optimal R/in. at lb/ft^3
Fiberglass	2.6–4.2	0.46–0.74	3.2
Cellulose	3.2–3.6	0.56–0.63	1.0–2.0
Mineral wool	2.6–4.4	0.46–0.77	4.0
Vermiculite/perlite	2.1–2.4	0.37–0.42	N/A
Expanded polystyrene	3.6–4.4	0.63–0.77	N/A
Extruded polystyrene	5.0	0.88	N/A
Polyisocyanurate board	5.6–7.6	0.99–1.34	N/A

For fibrous materials such as fiberglass, cellulose, and mineral wool, an optimal packing density achieves maximum R. R-values at lower and higher densities than the optimal value will reduce R-value.
Table created using data presented in Krigger and Dorsi (2009), with modification.

slow-conducting air, or by creating bubbles of air or other gases in foams. By properly installing high insulating materials in buildings, one can achieve a high composite *R*-value, which should greatly reduce the flow of heat to the outdoors during the winter heating season or indoors during the summer cooling season. This should also make it easier for heating and cooling systems to maintain the indoor set point temperature for a longer time, thereby reducing the cost to operate through reduced fuel needs. Sometimes, properly insulating a building can also reduce the size of heating system needed to keep the building at its desired set point.

The most commonly used building insulating materials include fiberglass, mineral wool, blown cellulose, expanded or extruded polystyrene foam, polyisocyanurate foam board, and vermiculite. Fiberglass and mineral wool can be applied as batts, boards, or blown into a cavity or attic floor space. The *R*-values of each of these materials are typically reported as *R*-value per in. (or per cm in SI) and achieve a maximum based on how densely packed the material is in the application. That is, a more densely packed cavity will not result in a higher *R*-value once it surpassed a certain optimal fill density. *R*-Values for these common insulation types are provided in Table 8.2.

8.2.2 WATER HEATING

As noted in Section 8.1, after space heating and cooling, water heating is the next single largest energy user in residential buildings in the OECD. According to the US DOE, domestic water heating is used primarily in showers and baths, clothes washing, dishwashing, and kitchen and bathroom faucets. In addition, according to IEA and EIA estimates, daily per capita hot water demand across the EU is 24 L compared to roughly 60 L in the United States and Canada (though highly variable due to climate variations across countries/regions). Commercial sector hot water demand is much lower globally, with the bulk of the demand originating in hotels, hospitals, and food service buildings.

Domestic water heaters are typically operated by transferring the heat from burning natural gas, propane, or fuel oil, or from resistive electric heating coils, to the domestic water supply. Demand is typically serviced via dedicated hot water heating systems or as part of a centralized space heating system. Dedicated systems come in different configurations, including storage or tank-based systems, "on-demand" or tankless systems, and increasingly, heat pump, solar, and geothermal energy systems. Fuels used for these heating systems in the OECD are mostly derived from gas and electricity, with an increasing demand for heat pump heating. A depiction of three popular types is shown in Fig. 8.15.

Hot water is generally not a survival need but is essential to quality of life in modern, affluent societies. Bathing, cleaning, laundry, and dishwashing all

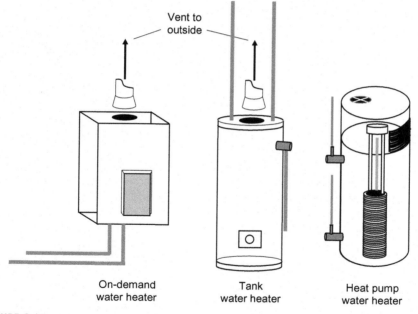

FIGURE 8.15

Three common types of water heating systems. On-demand water heaters typically use natural gas as the heating fuel and require venting. Tank water heaters typically use natural gas or electricity (heating coil inside, no venting required). Heat pump water heaters use electricity and require a condenser, an evaporator, a compressor, and a fan to operate. Cold water enters each system, which is heated and delivered to the area of the building where it is demanded.

Figure created using information and imagery presented in the DOE Energy Savers
Water Heating Infographic (2018).

benefit greatly from heated water. In the United States, for example, most water heating systems use a storage tank of 40–75 gal (151–284 L), with heating requirements to maintain the water at a high temperature, generally 120°F–140°F (49°C–60°C). This is especially true for households with tanks designed to hold enough hot water to fill any expected need. In these cases, the efficiency is a strong function of the insulation of the storage tank and its locations (inside a heated building as opposed to in a basement or a garage), in addition to the energy conversion and the heat transfer within the unit. Since a substantial amount of water is kept at a high temperature all day, every day, there is a considerable opportunity for heat loss during the majority of time when hot water is not being demanded. Especially if located in an unheated basement or garage, the driving force for heat loss may exceed 100°F during the winter.

According to the notable building science blog, Energy Vanguard, there are three ways energy is used and lost in water heating. The first, "firing energy"—or "demand energy" as noted by Krigger and Dorsi—is the energy used when converting a fuel, including electricity, to heat the incoming cold domestic water supply up to a desired set point temperature as water is used. In tank/storage water heater systems, water has already been heated and stored in an insulated tank, ready to be used as demand dictates. Any hot water pulled from the tank is replaced by cold water that must be heated back up to the set point. Tankless systems rapidly heat water only as needed but can be susceptible to not meeting multiple, simultaneous "calls" for hot water. Krigger and Dorsi note that firing or demand energy needs are a function of heating system efficiency, building occupant behavior, and consumption from fixtures (faucets and appliances). The second, "standby energy," is the energy that impacts only tank/storage water heaters. The large temperature gradient between the inside of a storage tank and a building's ambient temperature causes a constant loss of heat from the tank. This heat lost through the tank walls requires continual firing energy to maintain the set point temperature. The third, "distribution energy," is the energy that escapes through pipes and fixtures. The larger the distribution system, the larger these losses grow.

Given these known potential losses, it is then possible to rate water heating system efficiency. The commonly measured water heater efficiencies are:

- combustion or firing efficiency,
- recovery efficiency,
- energy factor, and
- overall system efficiency.

Combustion or firing efficiency varies by heating method and accounts for energy losses during combustion. A natural gas heater typically has a firing efficiency near or above 80%, while an electric heater has a firing efficiency of 100%, since the electricity is already delivered by the electric power plant (obviously, offsite power plant losses are not included). *Recovery efficiency* accounts for losses during the water heating process (energy needed to raise the temperature of water from its cold inlet value to the set point). *Energy factor*, the most commonly reported water heater efficiency metric, accounts for recovery efficiency and energy losses by a pilot light and losses through the storage tank (standby energy). According to Krigger and Dorsi, water heater energy factors are reported as decimal value between 0.5 and 1.0 and describes the fraction of heat input that stays in the water that leaves the storage tank. The US federal government requires a minimum energy factor, based on tank size and fuel type. For example, 2015 minimum energy factors for a 40 gallon (151 liter) tank system are 0.62 and 0.95 for natural gas and electric heaters, respectively. Note that for on-demand heaters that do not use a pilot light, recovery efficiency and energy factor are the same. *Overall system efficiency* accounts for all possible heating system losses, including the water heater and the distribution system. It basically gives a measurement of the actual delivered energy to the points of demand within the building. As in space heating, insulating hot water pipes throughout the building reduces distribution losses. A breakdown of all losses in a hypothetical water heating system is depicted in Fig. 8.16.

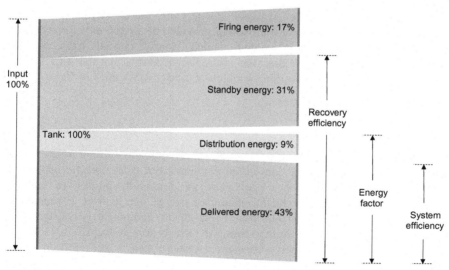

FIGURE 8.16

Typical heating losses and efficiency ratings for a hypothetical tank/storage domestic water heating system, using natural gas as a fuel. Note that overall system efficiency is an indicator of actual delivered energy to points of use and varies based on choice of fuel, whether the system is tank/storage or on-demand, and the size of the distribution system.

Figure created using composite information presented in Krigger and Dorsi (2009) and American Council for an Energy-Efficient Economy (2016).

As mentioned earlier, heat pump and solar hot water heating systems have gained increased popularity in recent years. Both improve overall system efficiency. Heat pump water heating systems operate under the same principle of heat pump space heating and cooling, but for water. They can be three times more efficient than a storage water system; however, heat pump performance depends on location, since they exhaust cold air, which increases heating needs in cold months. Solar hot water systems simply use panels on top of the building to preheat water, reducing the amount of heat needed from fuel to reach the high set point temperature. These systems can increase the efficiency of gas and electric heaters by 50%.

Finally, the United States also includes hot tubs and swimming pools as additional consumers of heated water. In these cases, the end use, the tub or pool, is actually a heat storage vessel. In-ground pools are insulated by the surrounding soil, but that soil is much more thermally conductive than air (and tends to be cool, even in the summer). Insulation blankets are commonly employed to mitigate heat loss through the surface of the water. Solar heating systems can also be employed to heat pools.

8.2.3 APPLIANCES

Appliance use and prevalence tend to be indicators for the development level of a country. Many major appliances ultimately provide energy services for refrigeration, cooking, and washing/cleaning, which clearly benefit from a well-established energy infrastructure. For example, according to the EIA's 2015 RECS, the major energy consuming appliances used in US homes include:

- refrigerators (including freezers),
- dishwashers, and
- clothes washers and dryers.

Though there other essential (and nonessential) appliances found in US households, the bulk of the energy consumption comes from these specific appliances listed above. In addition, since appliances are commonly desired to function promptly on demand, a significant amount of energy can also be consumed as "phantom loads," via internal electronics that ensure equipment is constantly on standby, drawing electricity while awaiting the user's next demand.

Refrigerators are large energy consumers because they must be left on all the time. As such, a great deal of emphasis has been placed on improving refrigerator efficiency. Refrigerator efficiency is typically measured by its *refrigerator energy factor*, which is a decimal value in units of volume cooled per kWh energy used in the refrigeration cycle per day. In the US market, for example, volume is measured in cubic feet and electrical energy is measured in kilowatt-hours. According to Krigger and Dorsi, the US Federal Standard in 1990 was 7.71 and the technical limit is 18.5. This translates to roughly 980–390 kWh of annual energy consumption, which manufacturers are fast approaching. On a percentage basis, the

Association of Home Appliance Manufacturers reports that since 1990, the energy consumption of an average refrigerator has fallen by more than 50%, and by 25% in just the last 10 years. Especially in the US market, these energy savings have greatly impacted residential electricity consumption, with refrigerators now only representing 7% of household electricity usage. The primary factors that have increased refrigerator energy efficiency include:

- better insulation and weather stripping,
- more effective temperature controls,
- higher efficiency compressors and motors, and
- better heat dissipation coils.

One interesting feature of efficiency improvements of refrigerator models is that increased efficiency has facilitated progressively larger capacity, thus negating some of the energy savings (also known as the "rebound effect"). In addition, model size and configuration affect energy consumption. Side-by-side refrigerator-freezer models use more energy than top or bottom loading freezers, due to surface area differences. Refrigerators that have built-in ice makers and that have auto defrost features also consume more energy. The US EPA *Energy Star* program has incentivized manufacturers to produce models that exceed Federal minimum efficiency standards. Finally, user-related behavior also impacts energy consumption. Freezers should be kept as full as possible and defrosted. Doors should also be opened as little as possible. Refrigerator and freezer coils should be cleaned annually.

Dishwashers are another major household energy consumer. As mentioned in the water heating section, dishwashers rely mostly on hot water drawn from domestic hot water systems, estimated to be between 80% and 90% of the total energy consumption for the device. An additional heating element is located inside the dishwasher, which is used for supplemental heating and drying (the supplementary water heating allows the domestic hot water set point to be reduced, thus saving energy). Also, reduced water needs for washing results in reduced energy needs from the water heater. Dishwasher efficiency is thus measured and regulated by its *dishwasher energy factor*, which is a decimal value in units of washing cycles per kWh (cycles/kWh) energy used (technically, dishwasher energy factor is the reciprocal of the machine electrical energy plus the water heating energy consumption per cycle).

According to Hoak, Parker, and Hermelink, of 466 models sold in the United States in 2008, dishwasher energy factor ranged between 0.47 and 1.14 cycles/kWh. The US Federal Standard is 0.46 cycles/kWh for standard size dishwashers, meaning a maximum annual dishwasher energy consumption of 467 kWh, assuming 215 cleaning cycles per year. 2016 *Energy Star* certified models have dishwasher energy factor ratings of at least 0.8, meaning they are at least 73% more efficient than the Federal standard. *Energy Star* ratings have continued to push increased energy efficiency: the 2016 certification rating was near 270 kWh. Factors that help improve energy efficiency in dishwashers include the use of sensors to improve washing efficiency and to maximize low and medium cycles, especially when dishwashers are not filled to capacity.

Clothes washers and dryers are the last of the major appliances that consume a significant amount of end-use energy. Gone are the days of hand-washed and line-dried laundry, especially in the United States, where the average family washes about 300 machine loads of laundry annually. Much like refrigerators and dishwashers, the energy (and water) efficiency of clothes washers and dryers have increased dramatically in the past 10–20 years.

As with dishwashers, clothes washers mostly rely on domestic hot water when needed, resulting in low direct energy consumption. Thus the efficiency of clothes washers sold in the market are evaluated using an *integrated modified energy factor* (IMEF), which, according to the DOE, is calculated as the clothes container capacity divided by the sum of the: (1) total weighted per-cycle hot water energy consumption; (2) total weighted per-cycle machine electrical energy consumption; (3) per-cycle energy consumption for removing moisture from a test load; and (4) per-cycle standby and off mode energy consumption. As of 2015, the minimum US Federal IMEF standard for top loading washers is 1.29 and 1.84 ft^3/kWh for front-loading washers (front-loading washers save nearly 50% of energy and water compared to top loading washers). Thus a minimally compliant 4 ft^3 front-loading washer used 300 times a year would require 652 kWh of energy annually. *Energy Star* standards have IMEFs of at least 2.06 and 2.38, respectively.

Clothes dryers can be powered using gas or electricity, with heat pump models also available. Efficiency for electric and gas models is measured using a *combined energy factor* (CEF), which the US DOE defines as "the quotient of the Federal testing load size divided by the sum of the machine equivalent electric energy use during standby and operational cycles, in units of pounds of clothes dried per kilowatt-hour (lb dried/kWh)." Most models sold in the United States are electric, and the current Federal standard CEF is 3.73 for electric dryers and 3.30 for gas dryers. According to *Energy Star*, the most energy-efficient standard size electric models (4 ft^3 capacity) in 2018 had CEFs of around 5.7, requiring only 189 kWh per year to operate, assuming 283 cycles per year.

According to *Energy Star*, choosing models that have sensors to monitor water temperature and moisture content helps reduce energy consumption. Also, choosing to use cold water for washing significantly reduces energy consumption in clothes washers, while using longer low-heat cycles for drying also reduces energy needs. Finally, in both machines, it is more efficient to wash and dry full loads.

8.2.4 LIGHTING

There are various needs for lighting inside and outside of buildings. In both residential and commercial buildings, there are needs for general overhead lighting, accent lighting, as well as lighting for specific tasks, such as cooking, office work, and machine work. Especially in urban areas, there is also an expectation for lighting outside buildings, and along streets and parks to provide a sense of security and better visibility after dusk. As such, lighting fixtures are ubiquitous in and around buildings.

At first glance, lighting appears to represent a relatively small share of total residential energy consumption in the OECD, on the order of 4%, and a moderate share of total commercial energy consumption in the United States at 10.4%. However, since lighting is an end-use that is dependent completely on electricity, the share of total *electricity* consumption from all lighting is closer to 20% in most OECD countries. In regions where space heating and cooling are not widely adopted or necessary, this percentage can be much higher. Because electricity continues to be generated primarily from burning fossil fuels, lighting represents an even more significant *primary* energy consumer; thus there are financial and environmental incentives to reduce energy consumption from lighting as discussed further in Chapter 9, Policy instruments to foster energy efficiency. In addition, as buildings become more energy efficient with respect to space heating and cooling, as well as water heating, the share of energy used for lighting will increase, further providing an incentive to reduce building-level lighting energy consumption.

Using the most recent US CBECS, which surveyed commercial building energy consumption in 2012, we find that the most commonly utilized types of lighting include:

- incandescent,
- halogen,
- standard tubular fluorescent,
- compact fluorescent lamp (CFL),
- high-intensity discharge (HID), and
- light-emitting diode (LED).

Each of these artificial lighting types has distinct characteristics and fulfill needs of a specific task. Because they are derived from different technologies, each has its own *color rendering index* (CRI), which defines an artificial light's ability to render correct color most closely to natural sunlight. Sometimes the ability to correctly render the color of natural light will make a certain lighting type more suitable for a task. In addition to color rendering, each lighting type has a technical lighting *efficacy*, which is defined as the amount of light output per unit of input electrical power, in units of lumens per watt, lm/W. Efficacy is the primary way we measure lighting efficiency.

Incandescent lamp technology has not changed significantly since its introduction in 1879. As we have noted in a previous work, incandescence typically involves the resistive heating of a tungsten filament to produce a white-hot glow. It is a very simple technology and inexpensive to produce. They also have the highest CRI: close to 100 out of a 100 scale. Incandescent lamps have been widely utilized since their creation; however, their primary drawback is that they are much more effective at generating heat than they are at producing useable light. That is, their efficacy is rather low, estimated to be on the order of 4−18 lm/W. According to Azevedo et al., when combined with a lighting fixture, efficacy reduces to 2−16 lm/W. Incandescent lamps also have relatively short service lifetimes and are best configured as small, round lamps, which do not offer the most effective means for ambient lighting in large indoor areas. Advantages are mostly tied to cost, high CRI, and being dimmable. Halogen lamps are a variant of incandescent lamp, which contain a mixture of inert gas and a halogen, such as bromine or iodine. This, combined with a heat-reflective coating inside the bulb, allows for the filament to remain hot using less electricity. Reported halogen device efficacy is 15−33 lm/W, and a device/fixture efficacy of 6−30 lm/W.

Fluorescent lighting technology has been, for many years, the chief competitor to incandescent lighting, primarily in the commercial sector, and more recently in the residential sector as well. The fluorescent light is produced from the excitation of mercury vapor via an arc generated between two electrodes. This excitation, in conjunction with a phosphor coating inside the fluorescent lamps, causes illumination. According to the Lighting Research Center, fluorescent lighting systems require the use of ballasts to regulate the current to the lamps and provide sufficient voltage to start the lamps. One of the drawbacks of this technology is its use of mercury, a known neurotoxin, which can be released if the lamp breaks.

As per Krigger and Dorsi, standard tubular fluorescent lamps are the most popular lamps for ambient lighting in large indoor areas, making it particularly suited for application in commercial buildings. Azevedo et al. report the efficacy of these long, narrow lamps to be 60−105 lm/W, which is much higher than the efficacy of incandescent lights. They also note that ballast efficiency can range from 65% to 95% and fixture efficiency can range from 40% to 90%, reducing the efficacy to 16−90 lm/W. Efficacy remains higher than that of incandescent, and lamp lifetime is much longer. The CRI for most "cool white" fluorescents ranges from 50 to 70, which has been a major complaint from users; however, recent technological advances have improved the color rendition significantly. Also, because of the excitation, older fluorescent tube lighting was plagued by a flickering that was sometimes a nuisance to the user.

CFLs are a second fluorescent lighting technology that has seen significant advances in the past 10 to 20 years, especially in the residential sector. Because of different lighting needs in residences, as well as the ubiquity of incandescent lighting fixtures in residential buildings, it was necessary to design a different ballast suitable for incandescent light fixtures. With the advent of electronic ballasts in particular, CFLs conform very closely to the size requirements needed to replace incandescent lamps in residential fixtures. Additional advances have also made CFLs more accepted, such as dimmable ballasts, and phosphor coatings that have significantly increased the CRI to levels close to incandescent. From an efficiency perspective, Azevedo et al. report CFL efficacy to be 35−80 lm/W, which, like tubular fluorescents, is significantly greater than incandescent. Ballast and fixture efficiencies are the same, reducing the efficacy with fixture to 9−68 lm/W.

HID lamps are a class of electric gas—discharge lamps, which operate under a similar principle to fluorescent lamps. Light is produced by a gas mixture being excited by an arc passing between two electrodes. The transparent arc tube is typically filled with a mix of an inert gas and a metal or metal salt. The tube produces a very hot, bright, high-pressure light output. Common HID lamps include high-pressure sodium, mercury vapor, and metal halide lamps. These high-pressure lamps require ballasts and are commonly suited to outdoor applications, and are popular choices in parking lots, and as road lights. They are also appropriate for certain indoor applications such as factories, gymnasiums, and warehouse-like retail stores. HIDs have a relatively long lifetime (up to 24,000 hours), are relatively small in size, and have CRI around 20. In terms of efficacy, Azevedo et al. report HIDs from 14 to 140 lm/W. They also note that ballast efficiency can range from 70% to 95% and fixture efficiency can range from 40% to 90%, reducing the efficacy to 4—120 lm/W.

LED lamps are a relatively new class of lamps, which have been disrupting the lighting market, due to their extreme flexibility and high efficacy. LEDs operate similarly to a photovoltaic cell, but in reverse. LEDs are known as solid-state devices, where an electric current is applied to a semiconductor diode, producing luminescence. Instead of ballasts, LEDs require *drivers*, which convert high voltage alternating current into low-voltage direct current suited to their proper operation. As a low energy, low-heat solid-state, high CRI device, LED lamps, and fixtures are being designed for every imaginable application, from residential ambient lighting to large area commercial building lighting, to street and parking area lamps. Azevedo et al. report current white LED efficacy from 60 to 188 lm/W. They also note that drive efficiency can range from 75% to 95% and fixture efficiency can range from 40% to 95%, reducing the efficacy to 18—170 lm/W.

In addition to efficacy, it is possible to estimate the percentage efficiency of a lamp's ability to convert electrical energy into useful light. According to Schubert, by definition, a monochromatic light source operating at an optical power of 1/683 W at 555 nm (the wavelength of light that the human eye perceives as brightest) has a luminous flux equal to 1 lm. This means that the maximum theoretical efficacy of a light source is equal to 683 lm/W. However, monochromatic green light is not a practical option for most lighting needs. White light, which covers a broader spectrum of wavelengths reduces this maximum theoretical efficacy down to 220 lm/W. So, conversion efficiency of a lamp can be compared to this definition as the ratio of lamp efficacy to the theoretical white light limit. In the case of an incandescent bulb, whose efficacy range is 4—18 lm/W, this would translate to roughly 1.8%—8.2% of the electrical energy input being converted to useful light. Including fixture losses, these numbers drop to 0.9%—7.3%. A table of efficacy and efficiency values for all discussed lighting technologies is provided in Table 8.3.

Based on these descriptions and the information provided in Table 8.3, it should be clear that the alternative options that now exist compared to the tried

Table 8.3 Lighting Efficacy and Electrical Energy to Light Conversion Efficiency for Commonly Used Lighting Technologies. Table based on a composite of data presented in Azevedo et al. (2009), LRC (2018), and DOE (2011)

Lamp Type	Ballast/Driver Efficiency (%)	Device Efficacy (lm/W)	Fixture Efficiency (%)	Combined Efficacy (lm/W)	Conversion Efficiency (%)	CRI	Rated Lifetime (h)
Incandescent	100	4–18	40–90	2–16	0.9–7.3	97	750–2500
Halogen	100	15–33	40–90	6–30	2.7–13.6	99	2000–3500
Tubular fluorescent	65–95	60–105	40–90	16–90	7.3–40.9	50–92	6000–20,000
CFL	65–95	35–80	40–90	9–68	4.1–30.9	82–92	10,000
HID	70–95	14–140	40–90	4–120	1.8–54.5	15–70	3000–29,000
LED	75–95	60–188	40–95	18–170	8.2–77.3	33–97	15,000–50,000

CFL, Compact fluorescent lamp; HID, high-intensity discharge; LED, light-emitting diode.

and true incandescent lamps of the past are reshaping how lighting is being used, particularly in the commercial sector. Moreover, because lighting technology that can reduce end-use energy intensity has advanced so quickly to meet new lighting efficiency standards, we can expect a significant reduction in energy consumption.

Reviewing data from the most recently available 2012 CBECS compared to the 2003 CBECS, the EIA reports that the most substantial change in lighting usage in the US commercial sector between those years was most closely related to "relamping" of incandescent bulbs with CFLs, as reproduced in Fig. 8.17. Tubular fluorescents had long been adopted by the commercial sector, but CFLs became the most logical alternative to incandescents, especially after new incentives and better CFL technologies came out during that 10-year window.

The EIA notes that lighting is the least cost prohibitive commercial energy-efficiency investment, and return on investment is most favorable. Indeed, according to the EIA, between 2003 and 2012, lighting's share of total US commercial sector

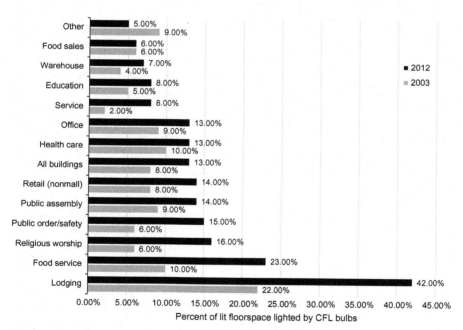

FIGURE 8.17

Difference between the percentage of lit floor space lighted by CFL bulbs in US commercial buildings in 2003 and 2012. A dramatic replacement of incandescent lamps with CFLs in most US buildings occurred during this time period. *CFL*, Compact fluorescent lamp.

Figure created using source data presented in the US EIA 2015 Residential Energy Consumption Survey.

electricity consumption dropped from being by far the largest at 38%, to just around 18%. This now places lighting at equal footing with refrigeration, ventilation, and cooling.

Finally, note that LED technology at the time of the 2012 CBECS was still only being installed by "early adopters" and market penetration was fairly nonexistent at that time. According to a 2016 National Electrical Manufacturers Association report, LED lamp market share increased from virtually zero in 2012 to a 17.1% share at the end of 2015. This should translate to even lower energy end use from lighting in commercial and residential sectors as more LED lighting technologies become more widely installed.

8.2.5 ELECTRONIC DEVICES

According to the US 2015 RECS, 20% of the electricity end use in residential buildings is tied to the various electronic devices that Americans have grown accustomed to owning. 7% of this total is represented by TVs, set top boxes, and Internet streaming devices, while computers and smart phones fall into the 13% "other" category, which also includes smaller kitchen devices (e.g., coffee makers, blenders, toasters). Indeed, electronic devices have becoming ubiquitous. According to the Consumer Electronics Association, whereas in 1980 the average US home had about 3 electronic devices, now it has 25.

Devices are major consumers of electricity in homes and in many commercial settings. Numerous items of relatively large electronic equipment are also prevalent in offices and increasingly in homes: computers, monitors, printers, copiers, TVs, and more. In order to facilitate rapid access to the services and saved user program settings, many of these devices routinely maintain some current to prevent waiting for warm-up. This standby usage is what constitutes a "phantom" current or load. Although these are very tiny loads individually, the sheer number of them adds up to produce a significant consumption of energy. Convenience creates one of the challenges to efficiency in this subsector of appliances and office equipment. A great deal of the phantom load is a direct result of serving the desire to have equipment instantly accessible, so that the equipment is constantly kept "on." Additional sources of phantom load are poor product design, specifically the use of less-efficient components and power sources.

Many of the devices that end users have access to in residential and commercial settings would benefit from being shut off when not in use; however, humans do not tend to make conscious rational decisions when convenience is in more abundance than scarcity of resources. Thus policies must be adopted to mandate or incentivize manufacturers to limit power needs of devices, especially for times when standby power prevails. An example of this approach is when the US EPA *Energy Star* program set out in the early 2000s to set stricter standby power standards on *Energy Star* certified electronic devices, specifying no more than 1 W of power be used when a device was on standby.

8.3 RESIDENTIAL AND COMMERCIAL ENERGY AUDITS

As was described in Chapter 6, Industrial sector energy efficiency, the energy audit is an important tool for reducing energy intensity, thus lowering fuel costs. Whereas the industrial sector focuses on increasing manufacturing process energy efficiency, the residential and commercial sectors focus mostly on whole-building performance and occupant behavior divided into the previously discussed categories:

- heating systems,
- cooling systems,
- water heating systems,
- transmission, including thermal losses, air leakage, and unwanted thermal gains,
- heating/cooling system distribution losses,
- lighting, and
- appliances.

Since building stocks can be in place for many years, it is likely that the building materials used and the practices followed during that time no longer conform with best materials and practices now. In addition, materials eventually degrade with time, which also affects building performance. These combined factors often lead to reduced human comfort and increased energy costs that occur when heating systems compensate for degraded or deficient building materials. Thus an energy audit is beneficial to uncover any underlying issues that are causing reduced building performance.

The professional residential/commercial energy auditor or auditor team will work with the building owner to accumulate necessary baseline information, including: (1) inventory of heating, cooling, and water heating equipment, and their energy usage in the form of bills over a year or 2-year time frame; (2) a site visit to do additional inventorying of lighting and appliance systems, including lamps currently installed in fixtures, as well the makes, models, and ages of all major appliances; (3) perform specific building performance testing, including the use of "blower doors," thermal imagers, and combustion testing to determine how "leaky" a building, if there are any notable "thermal breaks" in the structure, and if the heating systems are operating optimally; (4) interview the building owner to better understand occupant behavior, including the use of supplemental space heating; and (5) generate an audit report with recommended retrofit measures, including labor and material costs. A responsible auditor will also note all health and safety issues, including improper fuel venting and moisture problems.

As an example, let us consider what a general residential energy audit might find if performed in a cold-climate region such as in the Northeast United States. Fig. 8.18 depicts annual household site end-use energy consumption on a

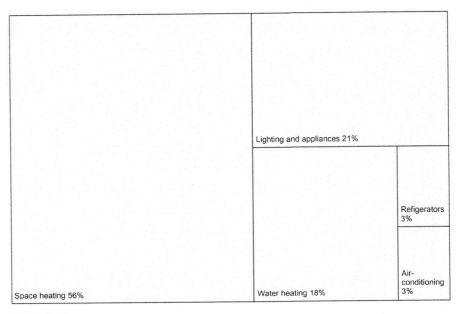

FIGURE 8.18

Annual household site end-use energy consumption on a percentage basis based on data collected in the 2015 US RECS for the total Northeast US housing stock.

Figure created using data presented in the US EIA 2015 Residential Energy Consumption Survey.

percentage basis based on data collected in the 2015 US RECS for the total Northeast US housing stock. As would be expected, space heating in this region dominates household energy end-use, while air-conditioning represents a very small portion. The housing stock in the Northeast United States is well established and older, which tends to promote inefficient building shells and heating systems with low seasonal efficiencies. As such, and as evidenced by the data presented in Fig. 8.18, space heating dominates the Northeast housing stock.

Assuming this breakdown of energy end use holds for an example New England home, an energy audit likely would find high air leakage, little or no insulation in the building walls, and an older, inefficient heating system. By performing basic air sealing, insulating wall cavities, and installing a new, more efficient heating system, with minimal distribution losses, it is realistic to reduce space heating needs by at least 40% in a poorly performing home. If these measures were carried out by the building owner, and the building achieved a 40% reduction in space heating needs, a new distribution of percentage energy end use would resemble something close to Fig. 8.19. By performing these energy-efficient measures, the example home now has a more evenly distributed end-use energy distribution. Note that in the postenergy retrofit, the contribution from lighting and appliances and from water heating are now

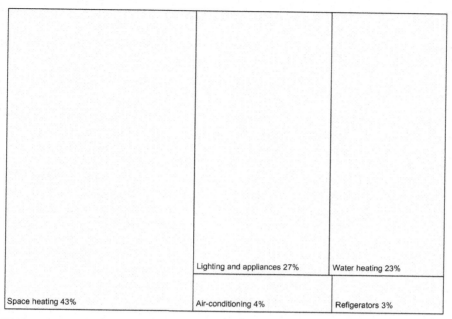

	Lighting and appliances 27%	Water heating 23%
	Air-conditioning 4%	Refigerators 3%
Space heating 43%		

FIGURE 8.19

Postenergy retrofit annual household site end-use energy consumption on a percentage basis for a Northeast US home, based on a 40% reduction in space heating energy end use due to improved air sealing, insulation, and heating system efficiency.

higher than in the preretrofit scenario. This does not imply they are now less efficient, but rather they just take a larger share of the total because of improved building performance and space heating efficiency that reduces the need for as much space heating. The same amount of fuel and/or electricity is still being used in these other systems. This a key point, especially with reference to the lighting efficiency improvements that were noted in the US commercial sector.

8.4 CALCULATION: HEAT LOSS THROUGH THE BUILDING SHELL WITH AND WITHOUT INSULATION

As we explained earlier, the ability of a building to perform the task of maintaining a set temperature using the least amount of fuel from the heating or cooling system is a key concern for energy efficiency. Insulation is a very effective, relatively low-cost strategy to reduce heat loss through the building shell, which, in turn reduces the amount of fuel needed to keep that set point temperature.

To estimate the effect insulation has on inevitable heat loss from a building, one can calculate the annual or seasonal heat loss through the building walls, as a function of the size or area of the walls, and the *R*-value of the walls. The basic equation for heat loss through a building shell is as follows:

$$\text{Building heat loss} = \frac{\text{Wall surface area} \times \text{temperature difference}}{R-\text{value}} \qquad (8.2)$$

Heat loss is typically denoted in the US customary system by units of Btu per hour (Btu/h), and in the SI by watts (W). Again, in the US customary system, *R*-value would be denoted by units of square feet (ft^2) times degrees Fahrenheit (°F) times hour divided by Btu (ft^2 °F h/Btu). Wall area, the surface area of the exterior walls, would be denoted by units of square feet (ft^2). Temperature difference, the difference in temperature between the presumably colder outside and warmer inside, would be denoted by "°F." Eq. (8.2) is essentially the hourly heat loss through the walls of a building. To calculate the *daily* heat loss for the building, one could record the temperature each hour of the day and add each heat loss value for the 24-hour period, or, to simplify the work, one can simply take the difference between the day's average outside temperature and the desired set point temperature inside the house and multiply that value by 24 hours. These equations, including units, are expressed as:

$$\text{Hourly building heat loss (Btu/h)} = \frac{\text{Wall surface area} \left(\text{ft}^2\right) \times \text{temperature difference (°F)}}{R-\text{value} \left(\text{ft}^2 \, °F \, h/Btu\right)}$$

$$(8.3)$$

and

$$\text{Daily building heat loss (Btu/day)} = \frac{\text{Wall surface area} \left(\text{ft}^2\right) \times \text{temperature difference (°F)}}{R-\text{value} \left(\text{ft}^2 \, °F \, h/Btu\right)}$$
$$\times 24 \, h/day$$

$$(8.4)$$

Note that building system heaters (furnaces, boilers, etc.) in the United States are typically rated and labeled by their hourly Btu output; thus the hourly heat loss can be crudely used as a proxy for properly sizing heating systems in residential buildings.

Calculating the *yearly* building heat loss is a little bit more complicated than just multiplying by 365 days, because wide temperature differences can occur between any 2 days, and, of course, between seasons. One could take the daily building heat loss and add the heat loss for all the days in the heating season. Fortunately, this is a commonly recorded value, known as *heating degree days* (HDD), which varies by climate zone in different geographic regions. This modifies Eq. (8.4) to be

$$\text{Seasonal heat loss} = \frac{\text{Building surface area}}{R-\text{value}} \times 24 \times \text{HDD} \qquad (8.5)$$

HDD are technically defined as the number of degrees that a day's average temperature is below 65°F (18°C), which is the temperature below which humans begin to feel thermal discomfort and buildings need to be heated. In the United States, HDD can vary significantly, with 2286 days in the West South Central region (e.g., Texas) and 6612 in New England (e.g., Maine).

Now with Eqs. (8.2)–(8.5), let us calculate the heating loss and influence of insulation on an uninsulated two-story, 1500 ft^2 home, which has roughly 5800 ft^2 of surface area located in the New England region of the United States, using oil to heat the home (20% of New England housing stock uses oil for home heating). Assuming an uninsulated wall R-value of about 4 and a 30°F temperature difference between outside and inside, the hourly building heating loss is equal to 5800 × 30/4 = 43,500 Btu/h. This means that the heating load of the heating system would have to operate at this heat loss rate to maintain the set point temperature (assuming 100% heating efficiency). Seasonal heating needs would be equal to 5800 × 24 × 6612/4 = 230 million Btu (242 GJ) for the season. That is equivalent to 1648 gallons of heating oil in a season, which is a significant amount of heating oil and a good candidate for adding insulation to the wall cavities.

After seeing this, the building owner wisely decides to install R-15 insulation into the wall cavities (a 3.5 in. blown-in product such as fiberglass or cellulose, or spray foam), which would provide an effective R-value of 19 (15 plus 4 from the wall's building materials). The building hourly heating loss is now reduced to 9167 Btu/h and seasonal heating needs reduce to 48 million Btu for the season. This is equivalent to 347 gal of heating oil in a season. At US$3 per gal of oil, the building owner would save US$3903 in one heating season, and if the entire insulation job costs $5000, simple payback would be 5000/3903 = 1.28 years. Even if it took US$10,000 to complete the job, payback, at this fuel price, still would be less than 3 years.

REFERENCES AND FURTHER READING

Books and Technical Articles/Reports

Azevedo, I. L., Morgan, M. G., & Morgan, F. (2009). The transition to solid-state lighting. *Proceedings of the IEEE, 97*(3), 481–510.

Bahadori, A., & Vuthaluru, H. B. (2010). A simple method for the estimation of thermal insulation thickness. *Applied Energy, 87*(2), 613–619.

Blok, K., & Nieuwlaar, E. (2016). *Introduction to energy analysis.* Taylor & Francis.

Cullen, J. M., & Allwood, J. M. (2010a). The efficient use of energy: Tracing the global flow of energy from fuel to service. *Energy Policy, 38*(1), 75–81.

Cullen, J. M., & Allwood, J. M. (2010b). Theoretical efficiency limits for energy conversion devices. *Energy, 35*(5), 2059–2069.

Cullen, J. M., Allwood, J. M., & Borgstein, E. H. (2011). Reducing energy demand: What are the practical limits? *Environmental Science & Technology, 45*(4), 1711–1718.

Ebenhack, B. W., & Martínez, D. M. (2013). *The path to more sustainable energy systems: How do we get there from here?* Momentum Press.

Hoak, D. E., Parker, D. S., Hermelink, A. H., & Center, F. S. E. (2008). How energy efficient are modern dishwashers? In: *Proceedings from the 2008 ACEEE summer study on energy efficiency in buildings* (Vol. 1).

Krigger, J., & Dorsi, C. (2009). *Residential energy: Cost savings and comfort for existing buildings*. Saturn Resource Management.

Parker, D. S., Fairey, P., & Lutz, J. D. (2015). Estimating daily domestic hot-water use in North American homes. *ASHRAE Transactions*, *121*(2), 258−271.

Romer, R. H. (1976). *Energy: An introduction to physics*. Freeman Press.

Zheng, X., Wei, C., Qin, P., Guo, J., Yu, Y., Song, F., & Chen, Z. (2014). Characteristics of residential energy consumption in China: Findings from a household survey. *Energy Policy*, *75*, 126−135.

Energy Information Sources and Technical Reports

American Physical Society. *Think efficiency: How America can look within to achieve energy security and reduce global warming*. (2008). Available from <https://www.aps.org/energyefficiencyreport/>.

Hart, R., & US Department of Energy Building Energy Codes Program. (2016). Introduction to commercial building HVAC systems and energy code requirements. In: *PNNL-SA-120201*. Available from <https://www.energycodes.gov/resource-center/training-courses/hvac-systems>.

International Energy Agency (IEA). *Energy efficiency indicators: Fundamentals on statistics*. (2014). Available from <https://webstore.iea.org/energy-efficiency-indicators-fundamentals-on-statistics>.

International Energy Agency (IEA). *Energy efficiency indicators: Highlights*. (2016). Available from <https://webstore.iea.org/energy-efficiency-indicators-highlights>.

International Energy Agency (IEA). *The IEA energy efficiency indicators database*. (2017). Available from <https://www.iea.org/newsroom/news/2017/december/the-iea-energy-efficiency-indicators-database.html>.

International Energy Agency, Energy Technology Systems Analysis Programme (ESTAP). *Technology brief R03: Water heating*. (2012). Available from <www.etsap.org>.

International Energy Agency, Tracking Clean Energy Progress (TCEP). *Heating in buildings*. (2018). Available from <https://www.iea.org/tcep/buildings/heating/>.

US Department of Energy Building Energy Data Exchange Specification. *Combined energy factor*. (2018). Available from <https://bedes.lbl.gov/bedes-online/combined-energy-factor>.

US Department of Energy Office of Energy Efficiency and Renewable Energy, ENERGY SAVER. *"2011 Buildings Energy Data Book" Table 5.6.9*. (2018a). Available from <https://buildingsdatabook.eren.doe.gov>.

US Department of Energy Office of Energy Efficiency and Renewable Energy, ENERGY SAVER. *Air conditioning*. (2018b). Available from <https://www.energy.gov/energysaver/home-cooling-systems/air-conditioning>.

US Department of Energy Office of Energy Efficiency and Renewable Energy, ENERGY SAVER. *Air-source heat pumps*. (2018c). Available from <https://www.energy.gov/energysaver/heat-pump-systems/air-source-heat-pumps>.

US Department of Energy Office of Energy Efficiency and Renewable Energy, ENERGY SAVER. *Energy saver 101: Home cooling infographic*. (2018d). Available from <https://www.energy.gov/articles/energy-saver-101-infographic-home-cooling>.

US Department of Energy Office of Energy Efficiency and Renewable Energy, ENERGY SAVER. *Energy saver 101: Water heating infographic*. (2018e). Available from <https://www.energy.gov/downloads/energy-saver-101-water-heating-infographic>.

US Department of Energy Office of Energy Efficiency and Renewable Energy, ENERGY SAVER. *Estimating costs and efficiency of storage, demand, and heat pump water heaters*. (2018f). Available from <https://www.energy.gov/energysaver/estimating-costs-and-efficiency-storage-demand-and-heat-pump-water-heaters>.

US Department of Energy Office of Energy Efficiency and Renewable Energy, ENERGY SAVER. *Furnaces and boilers*. (2018g). Available from <https://www.energy.gov/energysaver/home-heating-systems/furnaces-and-boilers>.

US Department of Energy Office of Energy Efficiency and Renewable Energy, ENERGY SAVER. *Heat pump water heaters*. (2018h). Available from <https://www.energy.gov/energysaver/water-heating/heat-pump-water-heaters>.

US Department of Energy Office of Energy Efficiency and Renewable Energy, ENERGY SAVER. *Maintaining your air conditioner*. (2018i). Available from <https://www.energy.gov/energysaver/maintaining-your-air-conditioner>.

US Department of Energy Office of Energy Efficiency and Renewable Energy, ENERGY SAVER. *Savings project: Lower water heater temperatures*. (2018j). Accessed from <https://www.energy.gov/energysaver/services/do-it-yourself-energy-savings-projects/savings-project-lower-water-heating>.

US Department of Energy Office of Energy Efficiency and Renewable Energy, ENERGY SAVER. *Sizing a new water heater*. (2018k). Available from <https://www.energy.gov/energysaver/water-heating/sizing-new-water-heater>.

US Department of Energy Office of Energy Efficiency and Renewable Energy, ENERGY SAVER. *Storage water heaters*. (2018l). Available from <https://www.energy.gov/energysaver/water-heating/storage-water-heaters>.

US Department of Energy Office of Energy Efficiency and Renewable Energy, ENERGY SAVER. *Wood and pellet heating*. (2018m). Available from <https://www.energy.gov/energysaver/home-heating-systems/wood-and-pellet-heating>.

US Energy Information Administration. *Gas furnace efficiency has large implications for residential natural gas use*. (2013). Available from <https://www.eia.gov/todayinenergy/detail.php?id = 14051>.

US Energy Information Administration. *Annual energy outlook 2015 with projections to 2040*. (2015). Available from <https://www.eia.gov/outlooks/aeo/pdf/0383(2015).pdf>.

US Energy Information Administration. *Trends in lighting in commercial buildings*. (2017). Available from <https://www.eia.gov/consumption/commercial/reports/2012/lighting/>.

US Energy Information Administration, TODAY IN ENERGY. *Dishwashers are among the least-used appliances in American homes*. (2017). Available from <https://www.eia.gov/todayinenergy/detail.php?id = 31692>.

US Energy Information Administration. *2012 commercial energy consumption survey*. (2018a). Available from <https://www.eia.gov/consumption/commercial/>.

US Energy Information Administration. *2015 Residential energy consumption survey*. (2018b). Available from <https://www.eia.gov/consumption/residential/>.

US Energy Information Administration. *What's new in how we use energy at home*. (2018c). Available from <https://www.eia.gov/consumption/residential/reports/2015/overview/pdf/whatsnew_home_energy_use.pdf>.

US Energy Information Administration, ENERGY EXPLAINED. *Energy use in commercial buildings.* (2018a). Available online at <https://www.eia.gov/energyexplained/index.php?page = us_energy_commercial#tab1>.

US Energy Information Administration, ENERGY EXPLAINED. *Degree days.* (2018b). Available from <https://www.eia.gov/energyexplained/index.php?page = about_degree_days>.

US Environmental Protection Agency, ENERGY STAR. *ENERGY STAR program requirements and criteria for dishwashers.* (2008). Available from <https://www.nepis.epa.gov>.

US Environmental Protection Agency, ENERGY STAR. *ENERGY STAR product retrospective: Standby power.* (2012). Available from <https://www.nepis.epa.gov>.

US Environmental Protection Agency, ENERGY STAR. *Lighting technologies: A guide to energy efficient illumination.* (2018a). Available from <https://www.energystar.gov>.

US Environmental Protection Agency, ENERGY STAR. *Clothes dryers.* (2018b). Available from <https://www.energystar.gov/products/appliances/clothes_dryers>.

US Environmental Protection Agency, ENERGY STAR. *Clothes washers.* (2018c). Available from <https://www.energystar.gov/products/appliances/clothes_washers>.

US Environmental Protection Agency, ENERGY STAR. *Dishwashers.* (2018d). Available from <https://www.energystar.gov/products/appliances/dishwashers/>.

US Environmental Protection Agency, ENERGY STAR. *ENERGY STAR most efficient 2018—Clothes dryers.* (2018e). Available from <https://www.energystar.gov/most-efficient/me-certified-clothes-dryers/>.

US Environmental Protection Agency, ENERGY STAR. *Refrigerators.* (2018f). Available from <https://www.energystar.gov/products/appliances/refrigerators>.

US Environmental Protection Agency, ENERGY STAR. *Room air conditioner buying guide.* (2018g). Available from <https://www.energystar.gov/products/heating_cooling/air_conditioning_room>.

US Federal Register. *Energy conservation program: Energy conservation standards for residential clothes washers.* (2012). Available from <https://www.federalregister.gov/documents/2012/05/31/2012-12320/energy-conservation-program-energy-conservation-standards-for-residential-clothes-washers>.

US National Weather Service. *What are 'Heating Degree Days' and 'Cooling Degree Days'?* (2018). Available from <https://www.weather.gov/ffc/degdays>.

Other Resources

American Council for an Energy-Efficient Economy. *Smarter house "Types of heating systems".* (2015a). Available from <https://smarterhouse.org/heating-systems/types-heating-systems>.

American Council for an Energy-Efficient Economy. *Smarter house "Replacing your water heater.".* (2015b). Available from <https://smarterhouse.org/water-heating/replacing-your-water-heater>.

American Council for an Energy-Efficient Economy, Smarter House. *Smarter house "Cooling systems.".* (2016). Available from <https://smarterhouse.org/home-systems-energy/cooling-systems>.

American Council for an Energy-Efficient Economy (ACEEE Blog). *Huge progress on lighting efficiency improvements.* (2016). Available from <https://aceee.org/blog/2016/05/huge-progress-lighting-efficiency>.

American Society of Plumbing Engineers. *Plumbing engineering design handbook, Volume 2. Domestic hot water systems.* (2014). Available from <https://www.aspe.org/sites/default/files/webfm/ContinuingEd/CEU_221_Mar15.pdf>.

Appliance Standards Awareness Project. *Product list.* (2017). Available from <https://appliance-standards.org/product>.

Association of Home Appliance Manufacturers. *Cool savings consumers to benefit from new refrigerator efficiency standards.* (2014). Available from <https://www.aham.org/AHAM/News/Latest_News/Cool_Savings__Consumers_To_Benefit_From_New_Refrigerator_Efficiency_Standards.aspx>.

Association of Home Appliance Manufacturers. *Cool energy savings: Guide to refrigerator and freezer energy comparison and cost savings.* (2018). Available from <http://coolenergysavings.org/>.

Bailes, A. (2011). Energy vanguard blog. In: *The 3 types of energy efficiency losses in water heating.* Available from <https://www.energyvanguard.com/blog/47612/The-3-Types-of-Energy-Efficiency-Losses-in-Water-Heating>.

Brenden, K. (2010). Why R is not simply the inverse of U. *Window and Door Magazine, July—August issue.* Available from <http://www.aamanet.org/upload/file/Why_R_Is_Not_Simply_The_Inverse_of_U_September.pdf>.

Lighting Research Center, Rensselaer Polytechnic Institute. *What is a ballast?* (2018). <https://www.lrc.rpi.edu/programs/nlpip/lightinganswers/adaptableballasts/ballast.asp>.

National Electrical Manufacturers Association. *LED A-line lamp shipments posted another strong quarter to close 2015.* (2016). Available from <https://www.nema.org/news/Pages/LED-A-Line-Lamp-Shipments-Posted-Another-Strong-Quarter-to-Close-2015.aspx>.

Pisupati, S. V. (2017). *John A. Dutton e-Education Institute course on energy conservation and environmental protection.* Pennsylvania State College of Earth and Mineral Sciences. Available from <https://www.e-education.psu.edu/egee102/l1.html>.

Shanley, S., & WegoWise Blog. *Calculate your building's heating energy intensity.* (2013). Available from <http://blog.wegowise.com/2013-03-07-calculate-your-buildings-heating-energy-intensity>.

Thermal Insulation Association of Canada. *Insulation materials and properties.* (2013). Available from <http://tiac.ca/wp-content/uploads/2015/12/TIAC_Guide_English_2013-Section-02.pdf>.

Policy instruments to foster energy efficiency

CHAPTER OUTLINE

Energy Efficiency. DOI: https://doi.org/10.1016/B978-0-12-812111-5.00004-4

Because the market has been unable to adequately reduce the efficiency gap, government intervention has played an important role. This chapter discusses the various policy instruments used to foster energy efficiency. Keypoints in this chapter include:

- understanding the various policy instruments available to foster energy efficiency,
- identifying the capabilities and limitations of government intervention in energy use, and
- examining a range of programs that exist to improve the efficiency of processes and products.

9.1 OVERVIEW

Public policy is commonly defined as purposeful government action in response to a policy problem, which is intended to modify or affect the behavior of firms, people, or subgovernments to solve the problem. In this case, the policy problem is the energy-efficiency gap, which was defined by Jaffe and Stavins as the difference between the current or expected future energy use and the optimal current or future energy use. That is, there is insufficient development, purchase, adoption, and use of energy-efficient actions, products, and technologies, which is a type of market failure.

Conceptually, market failure is an inefficient distribution of goods and services in a free market, which has multiple causes. The overarching policy objective to address the problem of the energy-efficiency gap is to use public policy to foster the development, purchase, adoption, and use of actions, technologies, and products that continuously increase energy efficiency. Achieving a policy objective typically means the identification and reduction of real or perceived barriers to addressing the problem (i.e., the root causes). Once identified, one must then formulate policies that are effective and cost-efficient in mandating, incentivizing, or educating firms or consumers to overcome identified barriers.

Because society is dynamic, the process of barrier identification and reduction and policy formulation is a continuous process. Likewise, because there are many energy-efficient technologies and products that are currently available and are constantly changing, in addition to those that have not yet been invented or commercialized, formulating policies to require or encourage energy efficiency is also a constant process. As a result, policies ideally need to be adaptable to address this dynamism.

Public policy formulation is a complex and often unpredictable process. This is a result of the interaction of a large number of governments, policy actors, industries, and nongovernmental organizations with diverse interests who all seek to influence the formulation process and policy outcomes. There is often disagreement as to the problem and/or the solution, which are influenced by national interests, political ideologies, culture, values, economics, resources, and environmental concerns. As a result of this complexity, public policy outcomes cannot be viewed as final actions, but steps toward achieving a policy objective, which also evolves.

As previously stated, public policy is purposeful government action (inaction can be purposeful action). "The government" is a commonly used monolithic term that describes multiple entities and actors given authority at multiple levels to govern a particular geographic area. As summarized in Table 9.1, there are various levels of government—multinational, national, state/provincial, county, and local (city/town)—which all have some capabilities and limitations to address energy-efficiency issues. Legal authority is provided for the respective jurisdictions dependent on the federal or state constitution and vested legislative powers for the executive and legislative branches at all levels. (The courts also play a key role in public policy as they interpret constitutions, laws, rules, and ordinances when there are gaps and inconsistencies and they rule on disputes and disagreements that can result in precedents to guide future policy decisions.)

In addition to constitutional and legal constraints, the relevant level of government is a function of the action to be addressed. A multinational or national government can set broad energy-efficiency policy objectives. For example, the European Commission's 2010 publication of its "20-20-20 by 2020" initiative established targets of a 20% reduction of greenhouse gas (GHG) emissions from 1990 levels, a 20% share of total energy consumption from renewable energy, and a 20% improvement in energy efficiency. This overarching policy objective was further refined with the 2012 adoption of the Energy Efficiency Directive that established more specific objectives through binding measures. The directive required energy distributors and energy sales companies to achieve a 1.5% energy savings per year through the implementation of energy-efficiency measures.

Table 9.1 Capabilities and Limitations of Governments With Energy-Efficiency Policy

Level	Capabilities	Limitations
Multinational and national	• Can establish uniform, broad-based minimum standards • Able to regulate interstate industries and commerce • Can influence the market as a large purchaser of goods and services • Can make substantial investments in research and development	• One-size fits all multinational or national approach can ignore regional and local needs and capabilities • Can intentionally or unintentionally stultify state/provincial policy innovation
State/province	• Able to address regional needs such as climate-specific building codes • Can control energy utilities • Able to influence the market as a medium to large purchaser of goods and services	• Limited financial ability to invest in research and development • Constrained ability to regulate international and interstate industries • Can intentionally or unintentionally stultify local policy innovation
Local county/city	• Best able to address local needs and constraints • Better suited to implement local transportation control measures	• Significantly strained financial resources • Small geographic influence • Can be limited by state or national policies

In contrast, a local government cannot impose efficiency standards on appliances, but they can write local building codes to promote energy efficiency in new construction. Likewise, a state government generally does not address the efficiency of a town's street lighting system, but they can establish minimum lightning efficiency requirements or provide grants to help local governments improve the efficiency of their street lighting. To be sure there are challenges as well. There is often a lack of concerted coordination within and among the various levels of governments and among national governments with regard to energy-efficiency policy.

Policies are developed within narrowed legal and political constraints based on real or perceived needs and capabilities. Unfortunately, policies are often formulated without sufficient regards to unintended consequences, which can be positive or negative, foreseen or unknown. For example, a classic case of a positive, foreseen consequence (co-benefitting) is policies that promote improved air quality, which generally necessitates increased energy efficiency and vice versa. Unrelated policies focusing on unrelated topics can also have unintended impacts on energy efficiency. For example, mandated safety features and equipment on vehicles (e.g., air bags, antilock brakes, stability control, tire-pressure-monitoring systems, etc.) can add significant weight to a motor vehicle thus reducing their efficiency.

However, even when carefully crafted policies are implemented to improve energy efficiency, the results are often lower than anticipated because of the so-called rebound effect. As defined, the rebound effect occurs when an increase in energy efficiency results in lower energy savings than anticipated because of behavioral responses. There are two perspectives as to the behavioral cause of the rebound effect. From an economics perspective, improved energy efficiency can result in a lower cost in operating a product or performing a service. If the demand remains stable, but the price decreases, there is more demand for the product or service. From a psychological perspective, when a person acts in a socially desirable way, such as consuming less energy, because of moral licensing, the person then feels free to act in a less socially desirable way in other facts (i.e., I have been good, now I can be bad). Common examples include the purchase of a fuel-efficient vehicle in which each mile of travel becomes cheaper resulting in the driver driving more miles, at faster speeds, or purchasing a larger vehicle all resulting in lower fuel savings. Another example is the installation of a low-flow shower head, the socially desirable action; the person may now take longer showers, the less socially desirable action.

This phenomenon means that when policies are designed, they must consider the potential for the rebound effect and measures need to be identified that can eliminate, reduce, or offset the effect. For example, as suggested by Vivanco, Kemp, and van der Voet, using proceeds from tradable allowance or carbon tax programs should be used to invest or to subsidize clean energy technologies and programs that further improve energy efficiency as approaches to offset the rebound effect.

9.2 POLICY APPROACHES TO FOSTER ENERGY EFFICIENCY

How specifically can the adoption of energy efficiency be fostered? Traditionally, we look to the market and the basic law of supply and demand to solve the problem. Intuitively, if there is a natural market demand for energy efficiency, then it will be supplied to a sufficient degree to meet the demand. Yet, it is much more complicated. One of the established dilemmas of the energy-efficiency gap is that consumers are reluctant to invest in energy-efficiency actions even though the investments would subsequently save them money. Simply providing information about potential energy savings or examples of energy savings has not been enough.

One of the explanatory reasons for this gap in consumer action in the world of electricity is the complexity to the average consumer of the language of electricity consumption—watts and kilowatt hours and appliance ratings, which is a type of information failure. For other energy consumption issues, such as motor vehicles, it is the use of miles per gallon, which does not necessarily resonate with the consumer. In addition, consumers tend to be biased on the purchase cost and do not adequately process operating costs even if the more energy-efficient product will result in lower operational costs compared to a non-energy-efficient product. Moreover, because of the volatility of energy prices, it is difficult for consumers to assess the potential payback.

As noted by Jaffe and Stavins, because of this uncertainly, consumers intuitively apply a high discount rate on future costs and benefits, which means that the payback has to be very high to justify, economically, the purchase price. Another reason is the split-incentive barrier, which is when the benefits are not realized by the party who pays for the product. For example, there is often no financial benefit for a building owner to purchase an energy-efficient appliance (assuming it costs more) if the tenant is the one who will benefit from the lower operating costs.

When there is insufficient or no market demand for energy efficiency in spite of the many social, environmental, economic, and national security benefits, public policy is warranted—public (government) intervention in the market to push or pull the demand and supply of energy efficiency. For example, in response to various countries' support of Israel in the 1973 Arab−Israel war, the Organization of Petroleum Exporting Countries (OPEC) imposed an oil embargo, which banned petroleum exports to targeted nations and introduced cuts in oil production. The result was a near tripling in the price of crude oil. In this case, government intervention by OPEC member states created a crisis in countries that were heavily reliant on petroleum necessitating government intervention to ameliorate domestic crises. The oil embargo also began the development of national energy strategies and subsequent adoption of public policies, which, in part, sought to reduce the consumption of petroleum through improved efficiency.

There are many policy instruments that can and have been used to foster energy efficiency, which is the focus of this chapter. The goal of this chapter is not to discuss every application of policy or to present arguments for or against each variation of a policy instrument, but to provide a focused discussion on the different categories and types of policy instruments that have been used for each of the major energy sectors: transportation, energy production, and commercial and residential consumers. As shown in Fig. 9.1, in this chapter, policy instruments are grouped into three major categories each relying on differing levels of government intervention and involvement[1]:

---

[1]Policy instruments are also sometimes divided into two major categories: hard policies (coercive) and soft policies (using attraction or cooption), which are sometimes referred to as "push and pull" policies, respectively. For example, a policy instrument that requires minimum efficiency standards seeks to push innovation while a policy of taxing inefficient vehicles seeks to pull demand for more efficient vehicles.

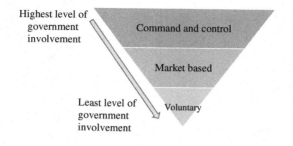

FIGURE 9.1

Three primary categories of policy instruments for energy efficiency.

- command and control,
- market based, and
- voluntary.

These categories should not be viewed as strict categories. Market-based policy instruments are often made possible or are backed up by command-and-control policy instruments and command-and-control policy instruments often seek some level of behavior change or voluntary participation. As shown in Table 9.2, there are a variety of policy instruments to achieve various outcomes related to energy efficiency that include tools, incentives, techniques, information, and actions, each with positive and negative attributes. As described by Karp and Gaulding, these categories utilize the respective motivational underpinnings with anticipation in achieving a certain behavioral or organizational outcome: command-and-control uses fear, market-based instruments use greed, and voluntary instruments capitalize on social responsibility.

9.3 COMMAND-AND-CONTROL POLICY INSTRUMENTS

As stated earlier, command and control is government rulemaking that prescribes allowed and prohibited actions, technologies, and/or outcomes that are imposed on a regulated entity. It relies on law, police power, and sanctions to compel an outcome. Command-and-control policy instruments, especially for numerical standards and specified technologies, provide a clear requirement to meet and are comparatively simple to monitor compliance. However, the rulemaking process is slow and can be unpredictable in the political arena, thereby creating future uncertainty as to the outcome of a rule.

Command-and-control policy instruments tend to be inflexible and fail to account for variability among firms, costs, and regions. Another criticism of the command-and-control approach is that there is no incentive to go beyond compliance such that the mandated maximum or minimum becomes the de facto

Table 9.2 Categorization and Summary of the Major Policy Instruments

	Policy Instrument	Summary	Example	Positive Attributes	Negative Attributes
Command and control	Mandated minimum efficiency standards	Mandate minimum efficiency standards for appliances, vehicles, and aircraft	Corporate Average Fuel Economy Standards in the United States	Outcomes based allows for flexibility and promotes innovation	The minimum often becomes the maximum
	Building codes	Mandate minimum requirements for efficiency for new buildings	National Energy Code of Canada for Buildings	Standards are generally clear, adoption of 3rd party standards by reference allows for continuous updates	Often covers only new buildings, which is a small number, increases purchase price
	Highway speed limits	Maximum speed limit based on efficiency	Former national speed limit of 55 mph in the United States	Easily enforceable, side benefit is reduced traffic deaths	Unpopular
	Mandated labeling and information disclosure	Specified energy information required on visible labels for consumers	Canada's EnerGuide program	Provide easy-to-understand comparative information	Modifying mandatory labels difficult and time-consuming
Market based	Taxes, tax credits, and tax deductions	Increasing or decreasing taxes to affect the purchase or operation costs of products or actions	Carbon tax	Greater flexibility, modifies behavior more efficiently, easy to comply	An unpopular approach at the consumer level
	Rebates, subsidies, and grants	Alternative mechanisms to reduce purchase costs for energy-efficient products	Efficiency Maine rebates for installing energy-efficient ductless heat pumps	Provides economic incentives to customers who may not otherwise purchase product	Not sufficient for lower income households, does not address the split-incentive problem
	Tradable allowances	Monetizes allowances for emissions that then can be bought, sold, or traded	The EU's Carbon Emissions trading program	Fosters innovation, generally least-cost approach	If priced too low, program will not work
	Time of use pricing	Uses pricing to promote shifting of demand to consume less energy during peak demand	The United Kingdom's Flextricity program	Consumers can still use energy; low-cost solution to modify the demand curve	Reduces consumer flexibility

Voluntary					
	Education	Increasing awareness and providing knowledge to be more efficient	Idaho Power Corporation's 30 simple ways to save energy	Relatively low cost to implement	Generally low effectiveness
	Default choice architecture	Nudging to socially desirable action using preferable default choice	Computer manufacturers select energy efficiency as default choice for when a computer enters standby	Voluntary, low cost, retains choice	Unintended consequences, success spotty
	Voluntary consumer labeling	Provide valuable energy consumption information for consumers to assess and compare	The Energy Star labeling program	Provide easy-to-understand comparative information, can foster innovation	Energy information is not among top factors for most consumers
	Technical assistance	Provides expert technical information and direct assistance	The Better Homes Alliance in the United States	Cost-effective approach for information diffusion	Small and medium communities and businesses generally underserved
	Industry challenges and voluntary agreements	Voluntary challenges to industry to develop or implement energy efficiency	US EPA's Green Lights program to switch to more efficient lighting	Low cost to administer, allows for maximum flexibility, not punitive	Generally limited in fostering innovation, the nonbinding nature of voluntary agreements reduces effectiveness

minimum or maximum, respectively. Thus while they can be very effective at achieving the specified outcome, command-and-control policy instruments are not generally cost effective with regard to achieving the desired outcome (i.e., they often are the most expensive approach to achieving an outcome).

The primary command-and-control-based policy instruments that have been used to foster energy efficiency include:

- mandated efficiency standards,
- building codes, and
- mandated labeling and information disclosure.

9.3.1 MANDATED MINIMUM EFFICIENCY STANDARDS

The goal of mandating efficiency standards is to intervene in the market and force the progress of technology while simultaneously removing inefficient technology. Without government intervention, consumers are often inclined to buy the lowest cost product on the market, which often means energy inefficiency. However, while the purchase cost may be low, the higher operation costs, due to higher energy demand, harm the consumer in the long run. Because the law of supply and demand often is most affected by short-term price fluctuations, it is difficult to develop technology that is likely to cost more, especially for major pieces of equipment (e.g., motor vehicles, airplanes, and appliances), which may take years to design and manufacture. Mandated minimum efficiency standards create a level playing field for manufacturers on a minimum technology and cost basis. In addition, the widespread adoption and diffusion of energy-efficient technology can drive down the cost due to economies of scale.

9.3.2 BUILDING CODES

Building codes are a set of requirements for the design and construction of a building. With energy efficiency, building codes typically address design and construction of the building envelope, ventilation, lighting, hot water, and heating and cooling systems. Codes are based on performance, achieving a certain level, or are proscriptive, such as specifying a height or thickness. Various nongovernmental organizations publish model codes in their respective areas, which are often adopted by state and local governments.

For example, the International Energy Conservation Code (IECC), published by the International Code Council (ICC), is an energy conservation code that establishes minimum requirements for energy-efficient buildings using prescriptive and performance-related provisions. The American Society of Heating, Refrigerating and Air Conditioning Engineers (ASHRAE) published its 90.1 Standards, which are minimum requirements for energy-efficient design and construction for new and major renovations of residential and commercial buildings.

In 2007, India's Ministry of Power launched the Energy Conservation Building Code that established minimum energy-efficiency requirements for new, large commercial buildings with a connected load of over 100 kWh, or a contract demand of over 120 kV amp. The building code includes requirements for the building envelope; lighting systems; heating, ventilation, and air conditioning (HVAC) system; and water heating. Canada adopted its National Energy Code of Canada for Buildings 2011, which established minimum energy performance level for buildings above three stories. Beginning in 1992, under the Building Codes Energy Program, the US Department of Energy (DOE) was charged with participating in developing model national codes in partnership with the ICC and ASHRAE. Since the inception of the Building Codes Energy Program in 1992, the cumulative energy savings from 1992 to 2012 are estimated to be approximately 4.2 quad.

9.3.3 MANDATED LABELING AND INFORMATION DISCLOSURE

As previously discussed, consumers are reluctant to invest in energy-efficiency actions even though the investments would save them money due in part to the lack of metrics meaningful to the consumer. To address the barrier of meaningful metrics, there are mandatory energy labeling requirements that apply a visual or graphical method designed to be simple to understand and reduce the time needed to compare costs for various models of the same product that seek to simplify consumer decision-making. When presented in these formats, energy savings are more relevant and less abstract to consumers. Moreover, labels by their nature create an incentive for manufacturers as their products can easily be compared to other products and encourages diffusion of efficiency technology among products and manufacturers.

9.4 MARKET-BASED POLICY INSTRUMENTS

Market-based policy instruments are tools that seek to encourage a desired behavior by modifying the economic incentive structure. In essence, market-based instruments seek to increase the economic cost of a socially undesirable action and/or lower the economic cost of a socially desirable action. As described by Karp and Gaulding, market-based policies try to capitalize on an individual's or firm's desire for economic reward as opposed to being compelled to act. Popular market-based instruments include taxes, tax credits and deductions, rebates, grants, tradable allowances, and time of use pricing. Market-based instruments allow for flexibility for the target population and can achieve the desired results at a lower cost compared to command-and-control approaches. Market-based instruments have weaknesses including the unpredictability of a market including external macroeconomic factors and the unpredictability of the response of participants (e.g., at what price signal does behavior change?).

The primary market-based policy instruments that have been used to foster energy efficiency include:

- taxes, tax credits, and tax deductions,
- rebates, subsidies, and grants,
- tradable allowances, and
- Time of use pricing.

9.4.1 TAXES, TAX CREDITS, AND TAX DEDUCTIONS

Research has shown that higher energy prices are associated with an increase in the adoption of energy-efficient products. However, energy prices are volatile, which means that if someone purchases an inefficient product during a period of low-energy prices, they are stuck with the product during a period of high energy prices. Therefore, relying on the market favors short-term decision-making as opposed to long-term efficiency goals.

The influence of supply and demand, by distorting (increasing) the price of, for example, energy, in theory will encourage the demand of products and services that consume less energy while discouraging the demand of products or services that are energy inefficient. The caveat is that any capital costs needed to purchase a product that is more energy efficient must have a sufficient and obvious payback period such that reduced energy costs will justify purchasing the product with a higher cost.

Taxes can be used to raise prices to the firm (manufacturer or importer) or to the consumer to influence demand and consumption. Taxes generally do not mandate a specific technology, action, or outcome, instead leaving firms with the flexibility to choose the least-cost method of achievement. Another important use of taxes is to encourage behavior change. An example is so-called sin taxes, which seek to discourage the purchase and use of tobacco products. To achieve behavior change, taxes are often required to be visible and separate and of sufficient cost to actually alter demand. That is, a separate tax that is charged at the retail point of sale where it is also recorded as a separate line item is more likely to change behavior than if the tax was simply embedded in the cost.

For example, after Ireland imposed a €0.15 tax on single-use plastic shopping bags at the point of retail, consumption dropped by 90% (Rucker, Nickerson, & Haugen, 2008). The separate, visible tax is designed to send a direct signal to the consumer of the monetary impact of a decision. A tax that is embedded into the price is often hidden from view and thus will not be obvious to the consumer. While the price may be higher, which could affect their decision, a visible and separate tax is designed to improve the likelihood of a behavior change.

With taxes designed to change behavior of the firm or consumer, revenue generation is, theoretically, not the primary objective, but revenue generation does occur. New or increased taxes generally are not politically or publically popular. One approach to improve their popularity is to ensure revenue neutrality—this is where other taxes are reduced at a rate equivalent to the rise in revenue from any new tax. For example, a new or increase in taxes on fuel could be offset by a corresponding decrease in taxes on the purchase of fuel-efficient products that rely on the taxed fuel.

One of the most significant barriers to installing energy-efficiency products is the capital costs, which can be substantial. To overcome this barrier, there have been and continue to be numerous national and state-level tax deductions and tax credits to lower the purchase cost, albeit after the purchase has been made.

9.4.2 REBATES, SUBSIDIES, AND GRANTS

Rebates, subsidies, and grants are all types of direct financial aid.[2] These strategies seek to directly or indirectly lower the purchase cost. This does not address the split-incentive problem, where, for example, a renter may pay for utilities, but the landlord owns the building and all systems. Rebates generally reduce or refund an amount that is applied to what has already been paid for a product or service. Rebates can be immediate, at the point of sale (similar to a coupon), or may be remitted following the completion of a formal rebate request. In contrast, subsidies are direct or indirect forms of financial aid from the government and often focus on reducing the cost.[3] (In the United States, energy conservation subsidies provided directly or indirectly to customers by public utilities are nontaxable, which further increases their financial worth.)

For example, low or no interest loans are a type of subsidy where the cost of the loan is decreased. Rebates and subsidies (and grants, which are discussed below) are important tools to overcome the barrier of capital (purchase costs). This is especially true with energy-efficiency actions where the greatest benefit is likely to be social rather than private. As reported by Girod et al., for example, subsidies (and labels) are the most effective policy instruments in the household sector to induce energy-efficiency innovation.

Monetary grants are a well-established instrument to implement policy. Monetary grants are nonrepayable funds provided by a government to a variety of organizations or sublevel governments for multiple purposes. With grants, governments can "hire" groups or sublevel governments to perform certain expressed functions in support of a policy goal or alteration of a behavior (e.g., providing grants to states or cities to adopt energy-efficiency programs). Grants can persuade recipients to perform actions they otherwise would or could not conduct themselves.

---

[2]See the Database of State Incentives for Renewables & Efficiency (http://www.dsireusa.org).
[3]See the DOE's Tax Credits, Rebates & Savings searchable database at https://www.energy.gov/savings/search.

9.4.3 TIME OF USE PRICING

Time of use pricing is a method that uses economic incentives to encourage consumers to shift their electricity demand to consume more of their electricity during times of low demand and thus less of their electricity during higher (partial peak) and peak demand periods. Although this approach has mostly been available to industrial and some large commercial customers that have the ability to be flexible with regard to energy consumption, the expansion of the smart grid allows residential customers to participate.

From an efficiency perspective, base-load power plants tend to be more efficient; older, less efficient plants (i.e., more expensive to operate) are typically brought online last. Thus, if demand can be shifted to reduce the high and peak demand curves, it decreases the need to use less efficient generation. Time use pricing sets energy prices for consumption during particular times of the day (peak times) or days (e.g., weekdays). In contrast, real-time-pricing is for the price of electricity for the period that it is consumed often updated on an hourly basis (sometimes even every 15 min) so that the consumer can make immediate decisions based on current demand. The latter option is likely to become more available with the increased expansion of the smart grid and smart meters.

9.4.4 TRADABLE ALLOWANCES

A tradable allowance program, referred to as "cap and trade," sets a limit on total, annual emissions. While the focus of a tradable allowance program may be to reduce a specific emission (e.g., CO_2), an indirect outcome is to improve the efficiency of the energy used because fuel consumption and emissions are closely linked. The goal of this instrument is twofold: (1) overtime, reduce emission caps to reach a predetermined level as shown in Fig. 9.2 and (2) use market forces to allow firms to make internal decisions on how best to reduce their allowable

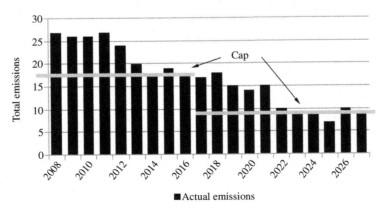

FIGURE 9.2

Hypothetical cap and trade program.

emissions. In essence, cap and trade uses the law of supply and demand by shrinking the supply of allowances to increase their value to encourage firms to make an internal economic decision as to the best approach to reduce emissions including increasing the efficiency of combustion.

Firms are given a certain number of allowances to emit the subject pollutant. At the end of the calendar year, firms must remit allowances to cover the previous year's emissions. If the firm does not have enough allowances, they can purchase or trade for unused allowances or use allowances the firm had previously banked. Or, the firm will have to pay a per allowance fine, presumably it will be higher than the purchase cost of unused allowances. As the cap shrinks, the number of allotted allowances also shrinks. In theory, it means that as the supply of allowances shrinks, and demand presumably stays the same or increases, the value of the allowance increases. Efficient firms will have excess allowances that they can sell or trade. In addition, the pollution control decisions would include energy efficiency as an option. That is, if a firm can produce the same amount of energy but by reducing fuel consumption, and less pollution is emitted, this will reduce the number of allowances needed.

9.5 VOLUNTARY POLICY INSTRUMENTS

Voluntary instruments are those that seek the cooperative efforts of the target population to achieve a certain desired goal. Voluntary-focused policy instruments rely more on social responsibility. However, they can be used to leverage action to avoid stronger government intervention. And, to be sure, individuals and firms will engage in voluntary actions if there are also some other benefits, particularly economic. They range from purely voluntary to legally binding agreements. Voluntary agreements can include self-regulation, negotiated agreements between a government and target population, voluntary partnership programs offered by government, challenge programs, information disclosure programs, and product energy labeling. The benefits of voluntary agreements is that they provide flexibility for the government and target population, can achieve the desired results at a lower cost, and can be relatively easily modified or ended. Voluntary actions can influence or obviate the need for government regulation. Voluntary agreements can be limited in their effectiveness because they are often nonbinding, and there is a free-rider problem where major players in the industry participate while others decline but may reap some benefit.

The primary voluntary-based policy instruments used to foster energy efficiency include:

- education,
- default choice architecture,
- voluntary consumer labeling,
- technical support, and
- industry challenges and voluntary agreements.

9.5.1 EDUCATION

Education focuses on changing consumer behaviors by providing clear and objective information that, if adopted, would result in improved energy efficiency. Often this is done through traditional education materials including brochures, posters, bill inserts, publishing of data, website portals, and social media. Energy-efficiency information has also been tailored to and adopted by schools for education curriculum. Education materials often adopt social marketing approaches. For example, focusing on the cost savings of improved insulation in buildings, the benefits of purchasing energy-efficient products, and how to access data from one's smart meters to increase energy efficiency and lower costs. A more active approach has been to provide free or reduced energy audits, which can provide specific information and action steps for a specific building.

9.5.2 DEFAULT CHOICE ARCHITECTURE

Default choice architecture is the careful and intentional design of choices such that the socially desirable choice confronted by the consumer is the default choice. With energy efficiency, the preferred default choice would be the most energy-efficient choice. The consumer could make a less energy-efficiency choice, but because it is not the default choice, they generally would have to make a concerted effort to make the less energy-efficient choice. For example, the energy-efficient default choice for a computer would be the standby and/or hibernation mode and the energy-efficient default choice in government and private procurement contracts would be to require the purchase of certified energy-efficient (e.g., Energy Star) computers.

9.5.3 VOLUNTARY CONSUMER LABELING

There are multiple programs (e.g., appliances, houses, and vehicles) that support the development and implementation of voluntary consumer labeling as a means to allow consumers to assess and compare products based on energy consumption. For example, Pacific Gas & Electric Company, an investor-owned utility in Northern California, labels certain products, such as lights, with a Smart Choice label, based on the product meeting criteria for energy efficiency and performance.

9.5.4 TECHNICAL SUPPORT

As a policy approach, the goal of technical support is to provide technical and economic data, technology, best practices, case studies, standards, and other issue-specific information that may not be easily accessible to some organizations (e.g., local governments, schools, residences, etc.). Technical support can include webinars, workshops, web-based information portal, newsletters, podcasts, and on-the-ground technical consulting.

9.5.5 INDUSTRY CHALLENGES AND VOLUNTARY AGREEMENTS

Challenges are a type of voluntary agreement where the government typically challenges an industry to voluntarily perform a particular function or achieve a desired outcome. (It should be noted that industries and trade groups often create their own energy challenge programs.) They are particularly useful where regulatory action is not appropriate or is highly problematic, such as adopting a particular, nascent technology or there is a likelihood of significant pushback. A side impact of voluntary challenge programs is the desire to promote technological innovation and also to increase adoption, thereby improving the economies of scale and thus lower the costs. Participants are also willing to spend their own money for challenge programs in part to "green" the organization. And, voluntary participation by an industry can reduce or obviate the need for regulation, which is nearly always more costly and restrictive.

9.6 POLICY INSTRUMENTS FOR THE TRANSPORTATION SECTOR

For the transportation sector, the policy objective is to reduce the consumption of fuel while performing the same level of work, which is done through a combination of programs that seek to discourage the purchase and use of fuel-inefficient motor vehicles, to reduce reliance on motor vehicles and especially single-occupancy motor vehicles, and to mandate efficiency standards for motor vehicles, commercial trucks, and airplanes. As shown in Fig. 9.3, the specific programs discussed in this section are as follows:

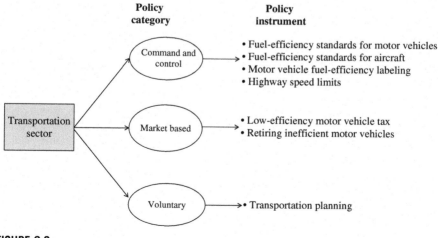

FIGURE 9.3

Primary policy instruments to foster energy efficiency in the transportation sector.

- fuel-efficiency standards for motor vehicles,
- fuel-efficiency standards for aircraft,
- motor vehicle fuel-efficiency labeling,
- highway speed limits,
- low-efficiency motor vehicle tax,
- retiring inefficient motor vehicles, and
- transportation planning.

9.6.1 FUEL-EFFICIENCY STANDARDS FOR MOTOR VEHICLES

In 1973, the average fuel efficiency of motor vehicles in the United States was 13.0 mpg (5.53 kpl). In 1975, the United States enacted the Energy Policy and Conservation Act of 1975. The act established Corporate Average Fuel Economy (CAFE) standards to push technology forward by mandating the minimum fuel efficiency of fleets of motor vehicles. (CAFE standards are set and enforced by the US Department of Transportation's National Highway Traffic and Safety Administration and the US EPA calculates the average fuel economy[4].)

Automobile manufacturers are required to meet a target for the sales-weighted fuel economy of its entire fleet of vehicles sold in the United States for each model year. The law mandated an increase in the average fuel economy of the new car fleet to 27.5 mpg by model year 1985. The 1978 model year was the first model year covered by the law, which required the CAFE standard for passenger cars to be 18 mpg and a separate CAFE standard for light trucks starting in model year 1979. Sport utility vehicles (SUVs), minivans, and crossovers were subsequently defined as light trucks subject to the lower CAFE standard. (Light trucks, SUVs, minivans, and crossovers account for about 65% of all new vehicle sales in the United States.) Based on a 2002 analysis by the US National Research Council, the CAFE standards were saving 2.8 million bbl per day, or 14% of total consumption.

The Energy Policy and Conservation Act of 1975 explicitly preempts states from setting their own fuel economy standards. However, under the Clean Air Act, while states are also preempted from setting their own vehicle emissions standards, California may establish its own vehicle emissions standards provided the US EPA determines the standards are necessary and they are as stringent as federal standards. This is because California's emission standards predated federal standards. If California receives a waiver under the Clean Air Act, other states may adopt the California standard, which currently includes 13 states. Thus, there are two sets of standards in the United States, California and Federal.

The minimum efficiency standards were not increased after 1985 because there was no explicit authority to modify passenger car standards until the passage of the Energy Independence and Security Act in 2007. From 1985 until 2011, the

---

[4]It should be noted that US EPA estimated mpg based on testing often overestimates actual MPG by about 20% because their number is based on laboratory testing and not real-world performance.

mandated standard for passenger cars was 27.5 mpg. While the light truck standards were increased to 20.7 mpg in 1996, they remained at that level until 2005. Regardless, fuel efficiency was increasing because of increasing oil prices and consumer demand.

The Energy Independence and Security Act of 2007 increased the CAFE standards for all passenger cars and light trucks combined to 35 mpg by 2020. In addition, the act mandated the study of the feasibility of developing fuel-efficiency standards for commercial medium-duty and heavy-duty vehicles.

In 2012 and subsequently in 2017, the National Highway Traffic Safety Administration (NHTSA) and US EPA published Phase 2 of the CAFE standards affecting model years 2017−25 light-duty vehicles (passenger cars, SUVs, vans, and pickup trucks), which was based on a multiparty agreement that included auto manufacturers, United Auto Workers union, and the State of California. The agreement was to increase the fleet-wide fuel economy to nearly 50 mpg by 2025.

In addition to light-duty vehicles, the Energy Independence and Security Act directed the NHSTA and US EPA to study the feasibility of increasing the efficiency of commercial medium- and heavy-duty vehicles with a gross vehicle weight rating (GVWR) of 8500 lb (3.86 t) or greater through test methods, measurement metrics, and fuel economy standards. The final rule established new standards beginning with model year 2021 for all heavy-duty vehicle and engine categories. A component of this program is the US EPA trailer standards (cargo trailers for tractor trailers), which went into effect for model year 2018 (NHSTA's standards take effect 2021). These efficiency-based performance standards encourage the use of aerodynamic devices, lightweight components, maximum tire-resistance standards, self-inflating tires, or a tire-pressure-monitoring system.

9.6.2 FUEL-EFFICIENCY STANDARDS FOR AIRCRAFT

The International Civil Aviation Organization (ICAO), a specialized agency of the United Nations to manage the administration and governance of the Convention on International Civil Aviation, was assigned jurisdiction over international aviation GHG emissions in the Kyoto Protocol in 1997. To address CO_2 emissions, an efficiency metric was developed for airplanes. In 2016, the ICAO finalized a performance standard, the first ever to impose binding energy-efficiency requirements that mandate fuel-efficiency improvements in fuel consumption of new aircraft delivered after 2028 using 2015 averages as the baseline. The performance standard requires an average 4% reduction with actual reductions ranging from 0% to 11% depending on the maximum takeoff mass of the aircraft. The efficiency standard is enforced under each National Aviation Authority's aircraft certification program; noncompliant aircraft would not be allowed to be certified or sold internationally.

9.6.3 MOTOR VEHICLE FUEL EFFICIENCY LABELING

Throughout the world, there are numerous mandatory fuel-efficiency labeling requirements for motor vehicles including Canada, Chile, China, Japan, Korea, New Zealand, Singapore, Thailand, and the United States. Since the mid-1970s, a standardized fuel economy label is required to be affixed to all new automobiles sold in the United States. Labels are required for seven different vehicle technologies: gasoline, diesel, ethanol flexible fuel vehicles, compressed natural gas vehicles, battery electric vehicles, fuel cell vehicles, and plug-in hybrid electric vehicles. Information required to be listed on the proscribed label includes estimated annual fuel cost; city, highway, and combined MPG; and fuel economy relative to other vehicles in the same class. The label also shows the estimated fuel cost over a 5-year period for the vehicle compared to the average new vehicle. Vehicles that run on liquid fuels display MPG and vehicles that run on other fuel types display gasoline-energy equivalent MPG.

9.6.4 HIGHWAY SPEED LIMITS

The goal of mandating speed limits is to decrease the consumption of fuel recognizing that motor vehicles have an optimum rate of efficiency at certain speeds. In direct response to the 1973 OPEC oil embargo in the United States, the Federal 1974 Emergency Highway Energy Conservation Act established the National Maximum Speed Law, which prohibited speed limits higher than 55 mph. (In 1942, the United States established the national maximum wartime speed limit, "Victory Speed Limit," of 35 mph to reduce gasoline consumption, which was eliminated in 1945.) To ensure adoption of the 55-mph speed limit by the states, for noncomplying states, the federal government could cut their federal funding for highway construction and repair. The National Maximum Speed Law was amended in 1987 and 1988 to allow speed limits up to 65 mph on certain roads. In 1995, Congress repealed the law allowing states to set their own maximum speed limits.

Research has shown that optimal speed for fuel efficiency depends on multiple factors, but especially engine size. Hooker (1988) found that of the single-occupancy cars tested, the efficiency penalty for driving at 75 mph compared to 55 mph was 31% for four-cylinder engines, was 28% for six-cylinder engines, and 33% for eight-cylinder engines. According to Meier and Morgan, a positive consequence of the speed limit was a sharp reduction in traffic deaths estimated to be 10,000 between 1973 and 1974 alone.

9.6.5 LOW-EFFICIENCY MOTOR VEHICLE TAX

In 1978, the US Congress created the Gas Guzzler Tax provisions in the Energy Tax Act of 1978 specifically to discourage the production and purchase of low-efficiency vehicles. However, the tax applies only to passenger cars; trucks, minivans, SUVs, and crossovers are not covered because these vehicle types were not widely available in 1978 and trucks were generally used for commercial purposes. Since 1991, cars with a combined fuel economy rating less than 22.5 mpg have been subject to the tax. The amount of the tax is based on a vehicle's combined fuel economy average and ranges from $1000 (21.5−22.5 mpg) to $7700 (less than 12.5 mpg), which covered 70 models in 2017. The tax, however, is collected from the vehicle's manufacturer and is typically included in the sales price although it appears as a line item on the car's sales sticker. Because trucks, minivans, SUVs, and crossovers are excluded, most of the so-called gas guzzlers are exempt from the tax.

In addition, the tax is applied to the fuel economy rating measured by the NHTSA. Yet, as reported by Tingwall, the fuel economy rating displayed on a vehicle's window sticker displays the US EPA's combined rating that is derived using a different methodology that can result in values 20%−25% lower than the NHTSA. Consequently, given the numerous exemptions to the tax and inclusion of the tax in the listed sales price, the tax has not been effective at promoting the purchase of more fuel-efficient vehicles.

Canada adopted a gas guzzler tax in 1989 (i.e., Green Levy Tax) on low-efficiency vehicles modeled after the US tax. Canada's tax, however, includes SUVs, vans with fewer than 10 seats, and crossovers, but not pickup trucks. The tax is based on the vehicle's fuel-efficiency ratings published by Natural Resources Canada and is applied to a weighted average fuel consumption of 13 L of fuel per 100 km or higher. The tax rate is between CAN$1000 and 4000 depending on the published fuel efficiency. Canada also imposed a CAN$100 tax on air-conditioners in motor vehicles because of their negative impact on fuel efficiency.

9.6.6 RETIRING INEFFICIENT MOTOR VEHICLES

One approach to improve the overall efficiency of vehicle fleets is vehicle retirement programs. Canada, France, Germany, Italy, Japan, Spain, and the United Kingdom have used this strategy. In the United States, the first federal vehicle retirement program was enacted in 2009, the Consumer Assistance to Recycle and Save program, commonly known as "Cash for Clunkers." This program, created in response to the Great Recession, was designed to be an economic stimulus program for the US automobile industry. In addition to the economic stimulus aspect, one of the goals of the program was to replace low-efficiency automobiles. Eligible consumers were provided vouchers of $3500 or $4500, which could be used to lower the price of a new car; the eligible trade-in vehicle had to be scrapped. The eligibility criteria include that the trade-in had to be less than 25 years old, the fuel economy rating was a combined 18 mpg or less, the vehicle must have been registered and insured continuously for the full-year prior to trade-in, and it had to be in drivable condition. New cars had to be less than $45,000. The rebate program was originally funded at $1 billion with an additional $2 billion added. The program ended in August 2009. Based on the program, the average fuel economy of the trade-in vehicles was 15.7 mpg compared to 24.9 mpg for the new replacement vehicle representing a 58.6% increase in fuel efficiency. However, this increase in fuel efficiency is not likely a sole result of the program as many consumers would likely have purchased a more fuel-efficient vehicle without the rebate.

California continues to operate a "cash for clunkers" car buyback program called the Consumer Assistance Program. The program's focus, however, is air quality improvement by providing cash incentives to retire older polluting cars. Eligible participants can receive a payment of $1000 ($1500 if the person is considered low income) if the car fails the state emissions test.

9.6.7 TRANSPORTATION PLANNING

Transportation planning focuses on the creation, operation, and management of facilities and services for various transportation modes for the efficient movement of people and goods. This involves, for example, the design and construction of roadways, parking, public transportation, and alternative transportation. Transportation control measures are strategies designed to reduce fuel use by decreasing vehicle miles traveled, vehicle idling, and improving traffic flow through careful managing of a region's or community's transportation system.

For example, between 1982 and 2014, the US DOE estimated 60.1 billion gallons of gasoline were wasted as a result of traffic congestion. Specific action related to energy efficiency on traffic congestion relief actions such as rotaries, improve traffic light signalization to reduce idling, removing traffic signals at intersections that no longer need signals, providing incentives to encourage carpooling or public transportation, eliminating or restricting drive-through services, and charging congestion tolls on roads.

The use of roadway and congestion pricing has recently increased. The focus is to use dynamic pricing that is correlated to the current level of congestion as a means to send a current price signal to drivers. The desired outcome is that drivers will shift driving to a less congested time to take advantage of a reduced toll charge or to shift to alternative transportation. Tolls are collected electronically to avoid toll-booth created congestion. In the United States, transponders are used; and in Germany, a GPS system is used.

To promote carpooling, generally defined as two or more passengers per vehicle, there are a variety of incentives to encourage the practice. These incentives (free parking, closer parking, ride matching, etc.) are sponsored by governments, businesses, and nongovernmental organizations. Programs may incorporate a variety of means to encourage employees to carpool. On a governmental level, the policy instrument most used to promote carpooling, where applicable, is through the use of dedicated lanes and reduced toll charges. For example, high-occupancy vehicle (HOV) lanes (defined as two or three or more occupants) are dedicated lanes reserved at peak travel times for the exclusive use of HOVs.

Because there are fewer vehicles and less congestion, this lane generally travels faster. For toll roads, HOVs are often treated differently. For example, in California and Virginia, HOVs are exempt from express lanes on toll highways.

Under the Federal Employees Clean Air Incentives Act of 1993 and the Transportation Equity Act for the 21st Century, US federal agencies in the National Capital Region (Washington, DC) implement a Transportation Subsidy Program. For federal employees who incur transportation expenses commuting to and from work, they are eligible to receive a transit pass for public transportation in the region. Regarding state-level policy to promote use of public transportation, in California, employers can offer transportation assistance with payroll tax savings because it is considered a tax-free benefit and not a taxable wage. Employers can subsidize a portion of their employees' commuting costs and allow employees to pay for the remainder of the costs with pretax dollars.

9.7 POLICIES FOR THE ENERGY PRODUCTION SECTOR

For the energy production sector, the policy objective is to produce and distribute energy more efficiently. As shown in Fig. 9.4, the specific programs discussed in this section are as follows:

- smart grid development,
- carbon tax,
- carbon emissions trading, and
- combined heat and power (CHP) partnership.

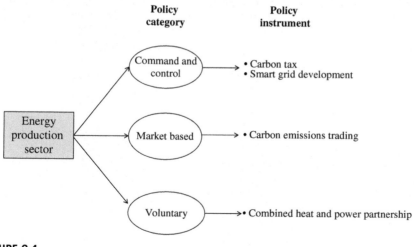

FIGURE 9.4

Primary policy instruments to foster energy efficiency in the energy production sector.

9.7.1 SMART GRID DEVELOPMENT

While the energy-efficiency benefits of smart grids are well established, the barrier to adoption and expansion is a result of the revenue structure of many public utilities. In the simplest terms, utilities that are regulated based on the cost-input model would need to invest substantial sums of capital (that would result in lower per customer revenues) unless ratepayers shoulder the costs, which is not popular. Changing the revenue structure to include incentives for efficiency could help reduce the barrier, similar to the United Kingdom's RIIO—Revenue = Incentives + Innovation + Outputs—framework for utilities.

The capital expenditure barrier with the traditional utility regulation model has led to the use of grants and subsidies to overcome the barrier of the capital expenditure, input model. For example, in the United States, the Public Utilities Regulatory Policies Act requires states to consider imposing certain requirements and authorizing certain expenditures pertaining to the smart grid. The DOE was provided funds under the American Recovery and Reinvestment Act of 2009 for the Smart Grid Investment Grant program to help fund cost-share projects with utilities.

In 2009, an EU Directive (DIRECTIVE 2009/72/EC) was passed that recommended the roll-out of smart metering systems across the EU but the Directive did not contain a binding target date for compliance. The European Commission subsequently specified 10 functionality requirements for smart grids, but final deployment is left to the Member States.

California passed a law to advance the smart grid. Some of the major objectives of the law included increased use of digital information and control optimization of grid operations and resources; development and incorporation of demand response, demand-side resources, and energy-efficient resources; deployment of cost-effective smart technologies, including real time, automated, interactive technologies that optimize the physical operation of appliances and consumer devices for metering; integration of cost-effective smart appliances and consumer devices; development of standards for communication and interoperability of appliances and equipment connected to the electric grid, and the identification and lowering of barriers to adoption of smart grid technologies and practices. Each of the state's three major investor-owned utilities are required to file Smart Grid Annual Reports to demonstrate progress.

9.7.2 CARBON TAX

A carbon tax is a tax on the carbon content of fossil fuels. The intent of a carbon tax is to change the relative price of carbon-based fuels by internalizing external social and environmental costs by penalizing the use of harmful fuels. The primary intent of carbon taxes is not to promote efficiency, per se, but to reduce carbon emissions related to global climate change by internalizing the externalized social costs. However, because of the carbon content aspect, a likely action by industrial consumers could be a shift to other fossil fuels with lower carbon content or alternative energy sources. An indirect outcome of reducing carbon emissions is to consume carbon-based fuels more efficiently.

For electrical generating plants, a carbon tax would increase the cost of producing electricity, which would be passed on to ratepayers. Thus, as the price of electricity increases, consumers of electricity would likely seek efficiency options to reduce their costs. According to the World Bank Group (2015), 17 national governments and one subnational government have enacted a carbon tax ranging from US\$1 to \$130 per tonne of CO_2 equivalent. Although many countries and states tax petroleum, this is not based on the carbon content, but a mechanism by which to raise revenues on a highly predictable consumable product.

A BTU tax based on the heat content of energy sources was proposed by US President Bill Clinton in February 1993 in part to promote energy efficiency, but the proposal was rejected by the US Congress. A BTU tax differs from a carbon tax because the BTU tax treats all fuels equally, as opposed to specific constituents and thus is more equitable regionally with regard to fuel-producing regions. The benefit would be an overall reduction in energy consumed and improved efficiencies as opposed to focusing on carbon-intensive energy sources and would reduce the incentive to simply switch fuels.

9.7.3 CARBON EMISSIONS TRADING

For the energy product sectors, tradable allowances programs (cap and trade) have been implemented to decrease SO_2 emissions to reduce acid precipitation and in various countries, regions, and states/provinces to reduce carbon emissions/GHG. While the primary goal is to significantly reduce total emissions using a market-based policy instrument, energy-efficiency projects are recognized as primary approach to attain low-cost reductions in GHG.

For example, the Regional Greenhouse Gas Initiative, a cooperative partnership among nine northeastern US states, was the first cap and trade program in the United States to establish a regional cap and reduction schedule of CO_2 emissions from the power generation sector. The partnering states sell nearly all emission allowances through auctions. The proceeds from the auctions are used in part to financially support energy-efficient actions and products in the private, public, and residential sectors. Based on a 2016 analysis by the Acadia Center, there was a 37% decrease in CO_2 emissions between 2008 and 2015 while electricity prices decreased by 3.4% primarily from switching to natural gas from coal (71% decrease) and oil (58% decrease). As a result of state investments in energy efficiency, the partnering states saved a cumulative total of 22,684 GWh of electricity since 2008.

9.7.4 COMBINED HEAT AND POWER PARTNERSHIP

The US EPA created the combined heat and power (CHP) Partnership, which is a voluntary program designed to promote efficient CHP technologies. US EPA offers various tools and services designed to facilitate and promote development of CHP projects. For example, the Partnership has a dedicated website, which includes tools, recommendations, and case studies of successful CHP projects; provides information state-level financial incentives for CHP; and a CHP emissions calculator to quantify pollution reductions from installing a CHP project at a particular site. Firms that become partner agree to assess the potential for CHP at their facilities and to publicize the various benefits of CHP.

9.8 POLICIES FOR THE COMMERCIAL AND RESIDENTIAL SECTORS

For the commercial and residential sector, the policy objective is to reduce the consumption of energy through a combination of programs that seek to discourage the consumption of energy and to prohibit and discourage the manufacture of inefficient appliances while simultaneously increasing the demand for energy-efficient products. As shown in Fig. 9.5, the specific programs discussed in this section are as follows:

- minimum efficiency standards—appliances,
- lighting efficiency standards,
- mandatory energy labeling,

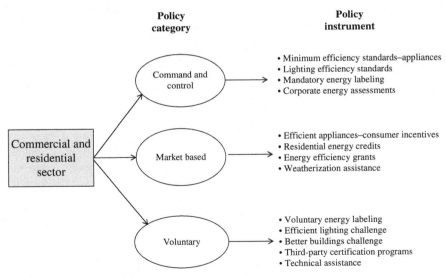

FIGURE 9.5

Primary policy instruments to foster energy efficiency in the commercial and residential sector.

- corporate energy assessments,
- efficient appliances, consumer incentives,
- residential energy credits,
- energy-efficiency grants,
- weatherization assistance,
- voluntary energy labeling,
- efficient lighting challenge,
- better buildings challenge,
- third-party certification programs, and
- technical assistance.

9.8.1 MINIMUM EFFICIENCY STANDARDS—APPLIANCES

Minimum efficiency standards have been adopted by multiple countries to eliminate inefficient appliances while spurring innovation to improve the efficiency of appliances. For example, in the United States, national minimum efficiency standards, originally mandated by the Energy Policy and Conservation Act of 1975, were promulgated for manufacturers of 33 specific categories of 60 consumer products representing about 90% of home energy use, 60% of commercial building energy use, and 30% of industrial energy use. These include refrigerators, freezers, dryers, dishwashers, refrigerators, furnaces, water heaters, lighted signs, microwaves, pool heaters, electric motors and pumps, and manufactured housing.

9.8.2 LIGHTING EFFICIENCY STANDARDS

In the United States, the Energy Independence and Security Act of 2007 mandated various efficiencies for lighting. The act required 27% greater efficiency by 2014 for common household light bulbs that traditionally consumed between 40 and 100 W. (The schedule of compliance was by wattage: 100-W bulbs by 2012, 75-W bulbs by 2013, and end with 60- and 40-W bulbs by 2014.) By 2020, the law also required that light bulbs be 60%—70% more efficient than the old, standard incandescent lamp. Specialty bulbs, three-way bulbs, chandelier bulbs, refrigerator bulbs, and plant grow lights were exempted from the efficiency requirement. In essence, it was a de facto ban on incandescent lamps between 40 and 100 W and such, is a command-and-control type regulation that mandates a performance standard, which is technology neutral. Thus, it does not mandate or prohibit any type of bulb, per se, but is designed to foster innovation.

A common adage in public policy is that budgets are policy. That is, where a government chooses to expend resources signals its priorities and interests. The US Congress in 2011 defunded the enforcement budget of the light bulb efficiency requirements as a de facto attempt to halt the program, but this failed to affect technological progress with the development and adoption of light bulb efficiency.

9.8.3 MANDATORY ENERGY LABELING

There are some 70 countries with mandatory energy labeling requirements. The EU's mandatory energy-efficiency labeling program for appliances requires labels to use an energy-efficiency scale of A, the most efficient, to G, the least efficient while simultaneously using a color code scheme with dark green, the most efficient, to red, the least efficient. The EU also requires manufacturers to notify consumers if software or firmware updates could reduce a product's energy efficiency.

Canada has the EnerGuide program, which is a mandatory energy consumption labeling program. Labels must contain the model's annual energy consumption in kWh, an energy consumption indicator that compares the model between the most efficient and least efficient models in the same class, and the type and capacity of models that makeup the class.

In the United States, EnergyGuide is the federal, consumer labeling program created in 1980. The program is designed to provide energy information displayed on a yellow label attached to selected products. (Products include clothes washers, refrigerators, freezers, televisions, water heaters, dishwashers, room air-conditioners, central air conditioners, furnaces, boilers, heat pumps and pool heaters, and ceiling fans.) The program allows consumers to compare estimated energy cost and energy consumption of different models within a given product category. The manufacturer must also post images of the EnergyGuide labels for products on a publicly available website. The program is administered through the DOE and the Federal Trade Commission. The Energy Labeling Rule requires manufacturers of certain appliances to disclose a product's annual energy cost or efficiency information and report their findings annually to the Federal Trade Commission.

The EnergyGuide label provides:

- estimate of annual electricity consumed by the product based on typical use,
- likely cost to operate the product based on national, average energy costs,
- list of key features of the product, and
- cost range for different models with similar features.

The EU also has mandatory energy labeling requirements for buildings. The Energy Performance of Buildings Directive introduced a framework for energy performance certification (EPC) for residential buildings. EPCs are required to contain details of the performance of the building (or building unit) by providing reference values for owners and tenants to compare the performance with other buildings. EPCs also contain recommendations for specific building improvements for cost-effective improvements. Estimates of cost savings resulting from improvements must also be included, and a forecast of underlying energy prices. An EPC is issued upon construction, sale, or rent of a building to the new owner or occupier.

9.8.4 CORPORATE ENERGY ASSESSMENTS

Under Australia's Energy Efficiency Opportunities Program, corporations that consume more than 0.5 PJ of energy in a financial year, and electricity generation companies, are required to conduct a corporate energy assessment. The energy-efficiency components of the assessment include the collection and analysis of energy consumption data, identifying and evaluating opportunities to improve energy productivity, and presenting the assessment outcomes to decision makers and the board as well as to the public. In 2013, the program was expanded to include corporations undertaking new developments and/or major expansion projects to help promote energy-efficiency planning. The program is supported with technical assistance including government experts, materials, and workshops.

9.8.5 EFFICIENT APPLIANCES—CONSUMER INCENTIVES

Manufacturers of specified energy-efficient dishwashers, clothes washers, and refrigerators can claim a business tax credit in the United States for each appliance manufactured. The tax credit is a sliding scale that is based on energy efficiency for refrigerators and both energy efficiency and water conservation for dishwashers and clothes washers. The tax credit ranges from $25 to $75 for each dishwasher, $175 to $225 for each clothes washer, and $150 to $200 for each refrigerator.

States and public utilities have a history of providing rebates to reduce the purchase cost of energy-efficient lights, appliances, and heating and cooling systems. For example, Efficiency Maine is the independent administrator for energy-efficiency programs in the US State of Maine. In the past, it provided rebates at retail check-out for efficient light emitting diode (LED) lighting. It now provides rebates for air sealing, insulating, ductless heat pumps, and high-efficiency central heating systems.

9.8.6 RESIDENTIAL ENERGY CREDITS

The US Energy Tax Act of 1978 authorized a federal tax credit for residential energy-efficiency investments for homes built after 1977; the program ended in the mid-1980s. The tax credit program was restarted with the Energy Policy Act of 2005, which authorized tax credits for hybrid vehicles, energy-efficient home improvements, and energy-efficient appliance manufacturers. As noted by Gillingham et al., numerous US states have offered tax credits and deductions including Arizona, California, Colorado, Hawaii, Montana, and Oregon while Arkansas, Idaho, and Indiana offered tax deductions during this same period.

9.8.7 ENERGY-EFFICIENCY GRANTS

DOE's State Energy Program (SEP) was created by the Energy Policy and Conservation Act of 1975. SEP provides federal financial assistance to the states in the form of annual formula grants and competitive awards. In 1990, the State Energy Efficiency Programs Improvement Act was enacted to encourage states, in part, to promote energy efficiency through public and private collaboration under the direction of DOE. The American Recovery and Reinvestment Act of 2009 was an economic stimulus spending package developed in response to the Great Recession of 2009. As part of the Recovery Act, $3.2 billion was provided for DOE's Energy Efficiency and Conservation Block Grant Program, which funded projects to improve energy efficiency and reduce energy use. In addition, $4.5 billion was authorized to convert buildings to high-performance green buildings, which seek in part sought to reduce energy use.

There have also been short-term and one-time federal grants for energy efficiency. For example, section 131 of The Energy Independence and Security Act of 2007 authorized the creation of a $400 million grant program to support transportation electrification. Funded projects focused on encouraging the use of plug-in hybrid vehicles.

9.8.8 WEATHERIZATION ASSISTANCE

Weatherization assistance programs focus on long-term reduction of energy consumption primarily for low-income households with financial support for one-time energy-efficiency upgrades to their residences. Participating households receive a free energy audit and a home retrofit that typically includes some combination of insulation, window replacements, furnace replacement, and infiltration reduction. In 2013, the United Kingdom launched its Green Deal program, which was terminated in 2015. A goal of the program was to increase the adoption of energy efficiency by retrofitting housing by using subsidies to reduce the capital costs. Households in England and Wales could receive a subsidy (voucher) of up to £1000 for installing two energy-saving improvements from a list of 12 eligible actions. Rebates were also available for up to 75% of the total cost of installing insulation for a maximum amount of £6000. In New Zealand, the Energywise Home Grants scheme was launched in 2004, followed by "Warm Up NZ" in 2009, which were grant programs focusing on insulating low-income homes.

9.8.9 VOLUNTARY ENERGY LABELING

Energy Star is a voluntary program for manufacturers to label the most energy-efficient models within a given category of products; generally the top 15%–35% efficient models. Although originated in the United States, the Energy Star label is recognized by Australia, Canada, Japan, New Zealand, Switzerland, Taiwan, and Europe. Manufacturers of qualified products may place the Energy Star label on qualifying products to indicate energy efficiency. Energy Star includes 37 household appliance and consumer electronic product categories including appliances, building products, commercial food service equipment, consumer electronics, heating and cooling, lighting, office equipment, water heaters, and miscellaneous (e.g., pool pumps, vending machines, water cooler, electric vehicle charging stations, and houses). For example, Energy Star houses consume a minimum of 15% less energy than those built to the 2009 IECC.

According to the US EPA, in 2015 alone, Americans purchased some 300 million Energy Star products resulting in a cumulative total of more than 5.5 billion Energy Star products since the program's inception. And, regarding homes, 82,000 new Energy Star rated homes were built in 2015 resulting in a cumulative total of more than 1.6 million certified homes.

9.8.10 EFFICIENT LIGHTING CHALLENGE

In 1991, the US EPA created the Green Lights Program, which was a type of challenge program to encourage the widespread adoption of energy-efficient lighting. To implement the voluntary program, the US EPA and a participant signed a memorandum of understanding that committed the participant to survey and upgrade their lighting system within 5 years. In the agreement, the US EPA offered a variety of programmatic support including an information portal, workshops, financial analysis tools, and so forth. According to the General Accountability Office, there were 2308 organizations participating in the Green Lights Program in 1997. These organizations had committed to perform lighting upgrades in 6 billion ft^2 of floor space equaling 9% of the national total.

In a different efficient lighting challenge, the US DOE launched the "L Prize" (the Bright Tomorrow Lighting Prize) competition designed to challenge industry to develop high-quality, high-efficiency solid-state lighting products. The first $10 million L Prize in the 60 W replacement category was awarded in 2011. Other prizes were established to replace the PAR 38 halogen bulb, which was suspended, and a third category, the 21st-century lamp (the requirements have not yet been published).

9.8.11 BETTER BUILDINGS CHALLENGE

DOE's Better Buildings Challenge challenges businesses, manufacturers, cities, states, universities, and school districts to commit to improve the energy efficiency of their buildings by at least 20% over 10 years. One of the goals of the program is to challenge participants (and to laud successful participants) in communicating energy-efficiency results though storytelling and achieving goals in media impressions, views, and online followers.

9.8.12 THIRD-PARTY CERTIFICATION PROGRAMS

Leadership in Energy and Environmental Design (LEED) is a program under the US Green Building Council, which serves as a third-party certification program for green buildings. It is a rating system on the sustainable performance standards of different kinds of buildings that covers the design, construction (new construction and major renovation), and operation of:

- data centers,
- health care,
- hospitality,
- retail,
- schools,
- warehouses and distribution centers,
- homes multifamily low rise (1−3 units), and
- multifamily midrise (4−8 units).

LEED uses a point-based system where projects earn LEED points for satisfying specific "green" building criteria. Certification is available in four progressive levels depending on the number of points obtained: platinum, gold, silver, and certified. The criteria include many factors, but one of the categories is Energy and Atmosphere, which contains 20% of all allocated points. This category consists of three prerequisites, which are mandatory to complete in order to obtain credits and sell credits. Points are awarded, for example, depending on the minimum energy cost savings percentage based on optimal energy performance. The latest version of LEED (v4) has also adopted smart grid technology as projects can receive points for participating in energy−demand response programs. India has a similar program, Green Rating for Integrated Habitat Assessment.

The Electronic Product Environmental Assessment Tool (EPEAT) is a voluntary global rating system for consumers to evaluate, compare, and select electronic products based on environmental factors. EPEAT is managed by the Green Electronics Council and was developed from a grant from the US EPA. EPEAT covered equipment includes:

- computers and displays,
- imaging equipment (scanners, copiers, printers, etc.),
- mobile phones,
- servers, and
- televisions.

To become registered under EPEAT, a product must meet environmental performance criteria including materials selection, design for product longevity, reuse and recycling, energy conservation, end-of-life management, and corporate performance. To obtain points in the energy category, the product must be rated as Energy Star under the US EPA and DOE. Using a point-based system, products are certified at different levels depending on the number of points earned: gold, silver, and bronze. EPEAT is often used in green purchasing for governments and businesses. For example, the federal government's acquisition regulations (e.g., 48 CFR 52.223-1) require the purchase of EPEAT-registered products.

9.8.13 TECHNICAL ASSISTANCE

DOE's SEP offers technical assistance to states to enhance energy efficiency. DOE created the State, Local, and Tribal Technical Assistance Gateway to provide an access point to its technical assistance and cooperative activities, which includes data, tools, and best practices. For example, the following are DOE technical assistance programs:

- *Better Buildings Alliance*—a collaborative program divided into sector teams, technology research teams, and campaigns focused on identifying and overcoming barriers to energy efficiency. The alliance works with associations, nonprofits, and trade organizations to host educational events. This has included supporting a voluntary national, commercial building certification program, and workforce training in energy efficiency.
- *Better Community Alliance*—a collaborative program similar to the Better Buildings Alliance but focuses on improving the prosperity of American communities through energy technologies.
- *Better Building Accelerators*—short-term, collaborative efforts that target specific barriers to energy efficiency including, for example, home energy information (increasing access and use of reliable and standard home energy information in real estate transactions), smart labs (promoting to improve energy efficiency in laboratory), and low-income housing support (identifies funding opportunities for expanding energy efficiency in low-to-moderate income communities).
- *State and Local Solution Center*—online source that provides resources to advance, high-impact energy-efficiency policies, programs, and projects.
- *Federal Energy Management Program*—provides federal agencies information, tools, project financing options, and assistance with energy-saving performance contracts to reduce the energy intensity of federal buildings.

9.9 CONCLUSION

As discussed in this chapter, there is a global energy-efficiency gap. A role of public policy is to identify real or perceived barriers to close the efficiency gap by identifying and adopting cost-effective policy instruments. However, the gap itself is not static because societies, behaviors, and technologies constantly change requiring policies to be adaptive. And, energy production and use are such complex and intertwined fields because there are multiple layers of government with national, regional, and local differences in energy needs and capabilities. Moreover, there are, to put it lightly, significant differences in opinion regarding the role of governments in intervening into markets regardless of the value of the outcomes.

As discussed, there are three major categories of policy instruments that rely on a descending need of government intervention: command-and-control, market-based, and voluntary. Each category has positive and negative attributes. Thus, no single policy is superior as it depends on the resources, players, and policy objectives. Nonetheless, as explored in the chapter, there have been numerous policy instruments that directly and indirectly have focused on improving energy efficiency with varying degrees of success. Finally, in developing energy-efficiency policies, one must always consider the likelihood of the rebound effect and to address ways to minimize or offset its impact.

REFERENCES AND FURTHER READING

Acadia Center. (2016). *Regional greenhouse gas initiative status report, Part I: Measuring success.* Retrieved from <https://www.ieca-us.com/wp-content/uploads/Regional-GHG-Gas-Initiative-Status-Report_Part-1_07.2016.pdf>.

DOE. (2014). *Building Energy Codes Program: National benefits assessment, 1992–2040.* PNNL-22610 Rev 1. US Department of Energy.

DOE. (2015). *Fact #897 November 2, 2015, Fuel wasted in traffic congestion.* US Department of Energy. Retrieved from <https://www.energy.gov/eere/vehicles/fact-897-november-2-2015-fuel-wasted-traffic-congestion>.

Gillingham, K., Newell, R., & Palmer, K. (2006). Energy efficiency policies: A retrospective examination. *Annual Review of Environment and Resources, 31,* 161–192.

Girod, B., Stucki, T., & Woerter, M. (2017). How do policies for efficient energy use in the household sector induce energy-efficiency innovation? An evaluation of European countries. *Energy Policy, 103,* 223–237.

Hooker, J. N. (1988). Optimal driving for single-vehicle fuel economy. *Transportation Research Part A: General, 22*(3), 183–201.

Jaffe, A. B., & Stavins, R. N. (1994). The energy-efficiency gap What does it mean? *Energy Policy, 22*(10), 804–810.

Karp, D. R., & Gaulding, C. L. (1995). Motivational underpinnings of command-and-control, market-based, and voluntarist environmental policies. *Human Relations, 48*(5), 439–465.

Meier, K. J., & Morgan, D. R. (1981). Speed kills: A longitudinal analysis of traffic fatalities and the 55 mph speed limit. *Review of Policy Research, 1*(1), 157–167.

National Research Council. (2002). *Effectiveness and impact of corporate average fuel economy (CAFE) standards.* Washington, DC: National Academies Press.

Rucker, R. R., Nickerson, P. H., Haugen, M. P. 2008. *Analysis of the Seattle Bag Tax and Foam Ban Proposal.* Northwest Economic Policy Seminar. Retrieved from <http://www.seattlebagtax.org/RuckerReport.pdf>.

Tingwell, E. (2014). Sting of guzzler tax, frozen for decades, fades. *New York Times,* September 13, p. AU2.

US EPA. (2016). *Energy Star® overview of 2015 achievements*. US Environmental Protection Agency. Retrieved from <https://www.energystar.gov/sites/default/files/asset/document/ES_OverviewAchievements_040816-508.pdf>.

US General Accountability Office. (1997). *Global warming: Information on the results of four of EPA's voluntary climate change programs*. Washington, DC (RCED-97-163).

Vivanco, D. F., Kemp, R., & van der Voet, E. (2016). How to deal with the rebound effect? A policy-oriented approach. *Energy Policy*, *94*, 114–125.

World Bank Group. (2015). *State and trends of carbon pricing 2015*. Washington, DC: World Bank Publications.

Index